Spatial Network Big Databases

KwangSoo Yang · Shashi Shekhar

Spatial Network Big Databases

Queries and Storage Methods

 Springer

KwangSoo Yang
Computer Science Department
Florida Atlantic University
Boca Raton, FL
USA

Shashi Shekhar
Department of Computer Science
 and Engineering
University of Minnesota
Minneapolis, MN
USA

ISBN 978-3-319-85964-4   ISBN 978-3-319-56657-3   (eBook)
DOI 10.1007/978-3-319-56657-3

Printed on acid-free paper

This Springer imprint is published by Springer Nature
The registered company is Springer International Publishing AG
The registered company address is: Gewerbestrasse 11, 6330 Cham, Switzerland

Preface

Spatial Network Big Data (SNBD) refers to spatial network datasets whose size, variety, or update rate exceeds the capacity of commonly used spatial network computing and spatial network database technologies to learn, manage, and process with reasonable effort. SNBD has the potential to transform society via next-generation routing services, emergency and disaster response, and discovery of potentially useful patterns embedded in these datasets. The use of SNBD is rapidly expanding into the transportation arena to improve the management and security of transportation infrastructure and enable data-driven decision-making. However, there are significant challenges to the use of SNBD because current methods, models, and algorithms do not always scale and/or perform well when storing, managing, and analyzing large volumes of data. Interestingly, most of the SNBD collected today are not used at all, and data that are used are not fully exploited. In addition, SNBD datasets tend to be used mostly for real-time control or anomaly detection, rather than optimization or prediction based on historical data, which is where their greatest value lies. Investigating novel SNBD storage and processing platforms, such as database systems and data analytics tools, is critical to realize the full value of SNBD.

Developing Spatial Network Big Database Systems (SNBDS) requires overcoming three key challenges. First, it requires new data models to represent the complex and interrelated structure of SNBD. Second, fully exploiting SNBD requires scalable query processing and optimization methods, which are currently lacking. Finally, SNBDS require I/O efficient storage and access methods that leverage scalability and efficiency of big data query processing. These challenges lead us to rethink both existing theories and models. This book presents a collection of concepts, algorithms, and techniques that effectively harness the power of SNBD. Reading this book is a first step toward understanding the immense challenges and novel applications of SNBD database systems. This book is organized in seven chapters. After reviewing some preliminaries in Chap. 1, we introduce basic graph algorithms in Chap. 2. Chapters 3–5 formally model spatial network query problems and explore algorithms that minimize the computational cost for query processing. Chapter 6 introduces strategies to develop I/O efficient storage and access methods. Chapter 7 summarizes the book's major themes.

Boca Raton, FL, USA   KwangSoo Yang
January 2017

Acknowledgements

I would like to express my sincere appreciation and thanks to my advisor Prof. Shashi Shekhar, Department of Computer Science, University of Minnesota. His enthusiasm, inspiration, and great efforts helped me to grow as a research scientist. I would like to thank Springer for their advice and support to publish this book. I would like to express my gratitude to the University of Minnesota Spatial Computing Research Group for their brilliant comments and suggestions. I would like to thank Kim Koffolt for improving the readability of this book. Lastly, I thank my mother, sister, brother, and the Lundberg family for everything they have done for me.

Acknowledgements

Contents

1 Spatial Network Big Databases: An Introduction 1
  1.1 Spatial Network Big Data 1
  1.2 Application Domain 2
  1.3 Spatial Network Big Database Management Systems 2
  1.4 Computational Challenges 6
  References 7

2 Basic Graph Algorithms 9
  2.1 A Brief Introduction to Graph Theory 9
  2.2 Network Representations 9
    2.2.1 Node-Node Adjacency Matrix 10
    2.2.2 Node-Edge Incidence Matrix 10
    2.2.3 Adjacency List 11
    2.2.4 Forward Star 12
  2.3 Shortest Paths 12
    2.3.1 Single-Source Shortest Path (SSSP) 13
    2.3.2 All-Pairs Shortest Paths (APSP) 14
  2.4 Block Decomposition 15
  2.5 Maximum Network Flow 16
    2.5.1 Augmenting-Path Algorithm 16
    2.5.2 Push-Relabel Algorithm 18
  2.6 Bipartite Weighted Matching 19
  2.7 Graph Partitioning 22
  References 24

3 Capacity Constrained Network Voronoi Diagrams 27
  3.1 Introduction 27
    3.1.1 Application Domains 27
    3.1.2 Problem Definition 28
    3.1.3 Problem Hardness 29

        3.1.4   Literature Review   30
        3.1.5   Outline of the Chapter   31
    3.2 Algorithms for Capacity Constrained Network Voronoi
        Diagram   32
        3.2.1   Pressure Equalizer (PE) Algorithm   32
        3.2.2   PE-BTCC Algorithm   35
        3.2.3   PE-Minor Algorithm   42
    3.3 Case Study with Brooklyn, NY Road Network   43
    3.4 Summary   44
    References   44

4 **Distance-Constrained k Spatial Sub-networks**   47
    4.1 Introduction   47
        4.1.1   Application Domain   47
        4.1.2   Problem Definition   48
        4.1.3   Problem Hardness   49
        4.1.4   Literature Review   49
        4.1.5   Outline of the Chapter   50
    4.2 Algorithm for Distance-Constrained k Spatial Sub-networks   50
    4.3 Case Study with Chicago Road Network   54
    4.4 Summary   54
    References   55

5 **Evacuation Route Planning**   57
    5.1 Introduction   57
        5.1.1   Application Domain   57
        5.1.2   Problem Definition   58
        5.1.3   Literature Review   58
        5.1.4   Outline of the Chapter   59
    5.2 Algorithms for Evacuation Route Planning   60
        5.2.1   Capacity Constrained Route Planner Algorithm   60
        5.2.2   Dartboard Network Cuts for Evacuation Route
                Planning Algorithm   62
    5.3 Experimental Analysis   67
        5.3.1   Experiment Design   68
        5.3.2   Experimental Results and Analysis   69
    5.4 Summary   71
    References   71

6 **Storage Schemes for Spatio-Temporal Network Datasets**   73
    6.1 Introduction   73
        6.1.1   Application Domains   73
        6.1.2   Basic Concepts   74
        6.1.3   Problem Statement   77

        6.1.4   Literature Review............................... 77

        6.1.5   Outline of the Chapter 78

  6.2   Lagrangian-Connectivity Partitioning Approaches for SSTN 79

        6.2.1   LCP-G1S for LGetOneSuccessor().................. 79

        6.2.2   LCP-G∀S for LGetAllSuccessors().................. 81

        6.2.3   Algorithm for LCP-G∀S.......................... 87

  6.3   Cost Models...................................... 90

  6.4   Experimental Analysis................................ 91

        6.4.1   Experimental Design 91

        6.4.2   Experimental Results and Analysis.................. 93

  6.5   Summary.. 96

  References 97

7 Summary... 99

  7.1   Capacity Constrained Network Voronoi Diagram 100

  7.2   Distance-Constrained *k* Spatial Sub-networks................ 100

  7.3   Evacuation Route Planning 100

  7.4   Storage Schemes for Spatio-Temporal Network Datasets........ 101

Chapter 1
Spatial Network Big Databases: An Introduction

1.1 Spatial Network Big Data

Increasingly, Spatial Network Big Data (SNBD) is of a size, variety, or update rate that exceeds the capacity of commonly-used spatial computing technologies to learn, manage, and process with reasonable effort [15]. Examples of SNBD include temporally detailed road maps that provide car speeds every minute for every road-segment, GPS trace data from cell phones, and engine measurements of fuel consumption, greenhouse gas emissions, etc. SNBD has the potential to transform our society. For example, a 2011 McKinsey Global Institute report estimates savings of about $600 billion annually by 2020 in terms of fuel and time saved by reducing vehicle congestion and idling at red lights or left turns [12]. WAZE and Uber are already easing congestion by offering alternative transportation. Scientists are investigating SNBD for hypothesis generation to address complex urban questions, where progress before was hampered by data paucity [17, 18].

However, there are significant challenges to the use of SNBD because current methods, models and algorithms do not always scale and/or perform well when storing, managing, and analyzing large volumes of data. Interestingly, most of the SNBD collected today are not used at all, and data that are used are not fully exploited [11]. In addition, SNBD datasets are used mostly for real-time control or anomaly detection, rather than optimization or prediction based on historical data, which provide the greatest value [11, 12]. Scalable and efficient SNBD storage and processing platforms, such as database systems, is critical to realize the full value of SNBD. However, developing Spatial Network Big Database Systems (SNBDS) requires overcoming three key challenges. First, it requires new data models to represent the complex and interrelated structure of SNBD. Second, fully exploiting SNBD requires scalable query processing and optimization methods, which are currently lacking [15]. Finally, SNBDS requires efficient I/O storage and access methods that leverage scalability and efficiency of large datasets [14]. These challenges lead us to rethink both existing theories and models. This chapter will explore key components at the conceptual, logical, and physical level of SNBDS to address these challenges.

© Springer International Publishing AG 2017
K. Yang and S. Shekhar, *Spatial Network Big Databases*,
DOI 10.1007/978-3-319-56657-3_1

1.2   Application Domain

Recent paradigm shifts caused by emerging autonomous car and car sharing technologies require new insight in the use of SNBD. If vehicles, sensors, and centralized traffic systems share important information with each other, then transportation infrastructures can use SNBD in a number of ways to improve traffic flow, avoid traffic accidents, save fuel, and reduce carbon emission on actual ridership (or rider behavior). The use of SNBD can also improve the utilization of assets such as parking space, transit service, etc. For example, public transit scheduling based on actual consumers and transit-tracking data could reduce commute or wait-time for buses and trains. In addition, transportation planners or engineers could use SNBD to identify and redesign road segments/stretches that pose risks for pedestrians. Indeed, SNBD can be used in a wide range of application domains from transportation to the electric power grid to water and gas distribution networks, etc.

1.3   Spatial Network Big Database Management Systems

This section provides an overview of three levels of SNBD database management systems. The levels form a three-level architecture comprising a conceptual, a logical, and an physical level, as shown in Table 1.1.

The conceptual level provides an SNBD conceptual data model, which helps us to define essential requirements and processes in the database design. An SNBD conceptual data model is a high-level description of the data requirements of the users (e.g., entity types, relationships, and constraints) and serves as a means for communication with domain experts, designers, and users. Examples of SNBD conceptual data models include Entity-Relationship Diagrams (ERD) and pictograms. Figure 1.1 shows an example of a pictogram that represents a spatio-temporal network.

The logical level transforms the high-level data model into an implementation data model. An SNBD logical data model provides a detailed data structure of a domain of information (e.g., tables, views, columns, primary keys, foreign keys, etc.). A logical representation is a crucial component of network analytics such as route evaluation, resource assignment, traffic flow analysis, and assessment of transportation resilience. The data model should support the development of new query

Table 1.1 Three levels of
SNBD database management
systems

Level	Sub-component
Conceptual	SNBD conceptual data model
Logical	SNBD logical data model
	SNBD query language
Physical	SNBD query processing strategy
	SNBD storage model

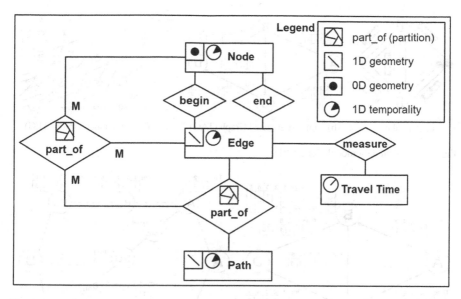

Fig. 1.1 Pictogram of a spatio-temporal network

languages and query processing for SNBD. Many transportation planning agencies
have begun collecting disaggregated spatio-temporal datasets (e.g., travel time, traffic
signal, congestion, etc.) at the user level in order to provide more detailed information
and better understand traffic patterns. SNBD datasets explicitly document multiple
time units to represent the dynamic nature of transportation networks. Tracking different time-periods allows us (1) to capture minute differences in travel times for
different routes, (2) to document associated traffic accident patterns, or (3) to identify increases in costs and demands for available resources over varying periods of
time. Prior work on SNBD models can be categorized into two groups: (1) Time-
Aggregated Graphs (TAG) and (2) Time-Expanded Graphs (TEG) [6, 19]. In both
models, a directed graph composed of nodes, edges, and temporal attributes represents SNBD. A TAG represents the changes in SNBD by collecting the node and
edge attributes into a set of time series [6]. Figure 1.2b illustrates a TAG for the road
network shown in Fig. 1.2a. In this example, every edge is associated with temporally
detailed travel times, as indicated by the array of numbers displayed alongside it.
Consider evaluating a route $(A \rightarrow B \rightarrow C \rightarrow D)$ at a time-step of 1 on the TAG.
Edge AB has a travel time of 1 at time = 1, edge BC has a travel time of 2 at time =
2, and edge *CD* has a travel time of 2 at time = 4. Therefore, the total travel time for
the route becomes 5. SNBD can also be modeled as a time-expanded graph (TEG),
which replicates each node along the time series such that a time-varying attribute is
represented between replicated nodes (see Fig. 1.2c) [19]. While TEG can explicitly
represent flow patterns without the need for elaborate procedures and algorithms, the
size of this model grows proportionally to the number of time-steps.

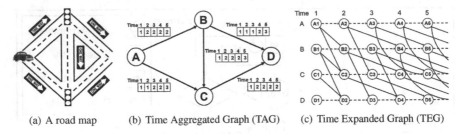

(a) A road map   (b) Time Aggregated Graph (TAG)   (c) Time Expanded Graph (TEG)

Fig. 1.2 Logical data models for SNBD

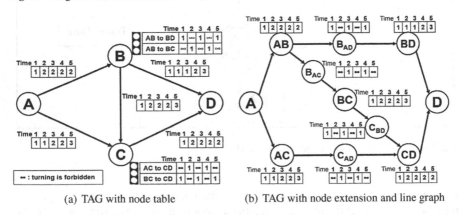

(a) TAG with node table   (b) TAG with node extension and line graph

Fig. 1.3 Logical data models with turn restriction constraints

However, both TAG and TEG pose a challenge in representing SNBD because of the variety of network constraints. For example, queries for enforcing turn-restrictions or lane-changes, reducing conflicts or crashes, avoiding obstacles, and reducing certain risk factors all involve complex relationships and significant constraints between different attributes. This complexity necessitates an efficient SNBD data structure to support various types of queries and data analysis. One feasible solution is to assign additional relations to nodes (or edge) in order to impose conditional constraints on the values of network attributes [8]. Figure 1.3a shows an example of a time-aggregated graph (TAG) where each edge is associated with travel times and each node in a traffic signal is associated with a relation for turn-costs. Consider, for example, evaluating a route $(A \to B \to C \to D)$. When starting at time = 1, it takes 1 time-step to reach B, 1 time-step to turn toward C, 2 time-steps to reach C, 1 time-step to turn toward D, and 1 time-step to reach D. However, the model outlined above shows a limitation to handling rapidly growing complex and interrelated datasets. For example, enforcing constraints for queries requires accessing multiple relations to get all the desired information, making it hard to ensure spatio-temporal topological consistency of SNBD. Whenever we transverse an edge, multiple relations are needed to evaluate a single route, leading to excessive join operations.

Table Name	Description
Node	create table node_table (node char, facility_name varchar(100), time-step time)
Edge	create table edge_table (node_a char, node_b char, distance int, capacity int, time-step time)

(a) Tables for nodes and edges

	Function	Example
Shortest Path	STN_dijkstra (text sql, char source, char target, time start-time)	STN_dijkstra ('select node_a, node_b, distance from edge_table', 'A', 'Z', 1)
Max Flow	STN_maxflow (text sql, char source, char target, text time-interval)	STN_maxflow ('select node_a , node_b, capacity from edge_table', 'A', 'Z', '1-9')
Min Cost Flow	STN_mcf (text sql, char source, char target, time start-time)	STN_mcf ('select node_a, node_b, capacity, distance from edge_table', 'A', 'Z', 1)
Network Voronoi Diagram	STN_nvd (text sql1, text sql2, time start-time)	STN_nvd ('select node_a, node_b, distance from edge_table', 'select node from node_table where facility_name='shelter'', 1)
Min-cut	STN_min_cut (text sql, text time-interval)	STN_min_cut ('select node_a, node_b from edge_table', '1-9')

(b) Spatio-Temporal Network Queries

Fig. 1.4 Example of SNBD query language

Node expansion and line graphs are one potential solution to address this challenge [8]. Figure 1.3b shows an example of time-aggregated graph (TAG) where every transitional point becomes a node associated with spatiotemporal attributes (e.g., turn cost and travel time) and every topological constraint becomes an edge to connect two transitional points. Consider, for example, evaluating a route $(A \rightarrow B \rightarrow C \rightarrow D)$. The evaluation can be conducted by following a series of nodes on the route: $A \rightarrow AB \rightarrow B_{AC} \rightarrow BC \rightarrow C_{BD} \rightarrow CD \rightarrow D$. This model can easily translate a query language into a general constraint optimization problem, such as shortest path, min-cost flow, min-cut partitioning, maximum matching, etc. However, the size of the network grows proportionally to the number of network constraints (e.g., spatio-temporal topological connectivity).

The logical level requires query languages that operate over spatial networks. Examples of query language include Structured Query Language (SQL) for relational databases and Object Query Language (OQL) and SQL3 for object databases [14]. However, these existing query languages do not fully and efficiently support spatial network queries (e.g., shortest path computation, network flow computation, Network Voronoi Diagrams, etc.). For example, although SQL uses a 'start with...connect by' clause to traverse graph datasets, it requires additional tables and complex inline queries to compute the spatio-temporal shortest path between two nodes. Figure 1.4 shows possible proposed examples of Spatio-temporal Network Queries.

A variety of approaches for network queries and optimization techniques have been developed to evaluate the best route for travelers and understand time-varying traffic patterns. For example, time-dependent routing problems incorporate a spatio-temporal network model into route computations and find the best routes in terms of travel cost [2, 3, 5, 7, 13]. Evacuation route planning problems extend a spatio-temporal network model to capacity-constrained flow optimization problems and analyze flow patterns during emergency situations [9, 10, 16]. K-nearest neighbor search problems assign a time-dependent weight on an edge and find the nearest service providers [1, 4]. Despite these advances, there are significant gaps in our understanding of network queries and it is helpful to consider multiple attributes and constraints on transportation networks to conceptualize real-world phenomenon.

The physical level transforms the implementation data model into an equivalent actual physical data model (e.g., database files, indexes, access paths, query processing, etc.). This physical level consists of two main sub-components: SNBD query processing strategies and SNBD storage models. The main goal of an SNBD query processing strategy is to minimize the execution time of spatial network data processing to answer a query. Examples of these strategies include network-traversal algorithms (e.g., breadth-first search, depth-first search, shortest path computation, etc.), network flow algorithms (e.g., maximum flow computation, minimum cost flow computation, etc.), and network partitioning algorithms (e.g., min-cut graph partitioning, Network-Voronoi Diagrams, etc.). An SNBD storage model is a storage scheme which describes data-structures, storage and access methods, and indexes. A storage model involves deep use of particular database management technology, such as data clustering and I/O efficient indexing. Examples of SNBD storage models include physical data models (e.g., node, edge, connection, etc.), index data-structures, network access methods (e.g., $getOneSuccessor()$, $getAllSuccessors()$, etc.) [19].

1.4 Computational Challenges

SNBD raises many computer science challenges. First, query processing methods are challenged by emerging use cases of spatial computing, such as route evaluation, flow analysis, resource allocation, and data analytics. SNBD requires new computational strategies for graph-based query processing. One of the biggest challenges in this problem is to minimize the computational cost to analyze SNBD and quickly respond to user requests. SNBD queries for network flow analysis are especially difficult with multiple attributes. Novel techniques are needed for pruning the search space in flow computations. In many cases, if complex and interrelated constraints make problems difficult then it is unlikely that efficient query processing mechanisms exist for SNBD queries.

In addition, the physical database design for an SNBD is challenged by temporal graph-based query semantics as well as the growing volume of temporally-detailed road-maps. Over the last few decades, researchers have tried to develop storage models for transportation networks that minimize disk I/Os in their respective fields by exploiting some structure of access patterns inherent in the datasets. However, the size of SNBD datasets themselves is so large (e.g., many TBs per day) that it is impractical today to even consider storing them on a single machine. Recently, big data processing platforms are a growing component for large-scale data management due to their potential ability to distribute large datasets across multiple machines and parallelize the query processing. To improve the accessibility and scalability for a database, the storage method not only needs to consider a cost model for each machine as well as the overall network, but also figure out data allocation rules that can efficiently parallelize the query processing in multiple machines to improve query response. A general rule for data allocation is that data should be placed as close as possible to where it will be used, and then load balancing should be considered to increase the global system performance. However, data allocation is challenging, especially when data grow in volume, and we need efficient incremental data allocation to process the real-time dataset.

This book explores these challenges via investigating scalable graph-based query processing strategies and I/O efficient storage and access methods. The chapters of this book are described briefly as follows. Chapter 2 reviews the basic network algorithms including shortest path, max flow, graph partitioning, etc. Chapter 3 introduces a special case of the Network Voronoi Diagram, namely the Capacity Constrained Network Voronoi Diagram (CCNVD). It describes the basic concept of CCNVD and presents methods that can create a CCNVD. It provides techniques to reduce the computation cost for creating a CCNVD. Chapter 4 introduces a special case of the network covering problem, namely, Distance-Constrained k Spatial Sub-Networks (DCSSN). It presents computational methods to create k sub-networks that can maximize the coverage of spatial events under distance constraints. Chapter 5 introduces the Evacuation Route Planning (ERP) problem and reviews current methods. It presents scalable methods to produce evacuation routes for large scale network datasets. Chapter 6 introduces the Storing Spatio-Temporal Networks (SSTN) problem and describes proposed approaches to store and access massive STN datasets. Finally, Chap. 7 summarizes the book's major themes.

References

1. Cruz LA, Nascimento MA, de Macêdo JA (2012) K-nearest neighbors queries in time-dependent road networks. J Inf Data Manag 3(3):211
2. Dean BC (2004) Shortest paths in fifo time-dependent networks: theory and algorithms. Rapport technique, Massachusetts Institute of Technology
3. Delling D, Sanders P, Schultes D, Wagner D (2009) Engineering route planning algorithms. In: Algorithmics of large and complex networks. Springer, pp 117–139
4. Demiryurek U, Banaei-Kashani F, Shahabi C (2010) Efficient k-nearest neighbor search in time-dependent spatial networks. In: International conference on database and expert systems applications. Springer, pp 432–449
5. George B, Sangho K (2012) Spatio-temporal networks: modeling and algorithms. Springer Science & Business Media
6. George B, Shekhar S (2008) Time-aggregated graphs for modeling spatio-temporal networks. In: Journal on Data Semantics XI. Springer, pp 191–212
7. Gunturi VM, Shekhar S, Yang K (2015) A critical-time-point approach to all-departure-time lagrangian shortest paths. IEEE Trans Knowl Data Eng 27(10):2591–2603
8. Hoel EG, Heng WL, Honeycutt D (2005) High performance multimodal networks. In: International symposium on spatial and temporal databases. Springer, pp 308–327
9. Kim S, Shekhar S, Min M (2008) Contraflow transportation network reconfiguration for evacuation route planning. IEEE Trans Knowl Data Eng 20(8):1115–1129
10. Lu Q, George B, Shekhar S (2007) Evacuation route planning: a case study in semantic computing. Int J Semant Comput 1(02):249–303
11. Manyika J (2015) The internet of things: mapping the value beyond the hype
12. Manyika J, Chui M, Brown B, Bughin J, Dobbs R, Roxburgh C, Byers AH, Institute MG (2011) Big data: the next frontier for innovation, competition, and productivity. McKinsey Global Institute

13. Orda A, Rom R (1990) Shortest-path and minimum-delay algorithms in networks with time-dependent edge-length. J ACM (JACM) 37(3):607–625
14. Shekhar S, Chawla S (2002) A tour of spatial databases
15. Shekhar S, Gunturi V, Evans MR, Yang K (2012) Spatial big-data challenges intersecting mobility and cloud computing. In: Proceedings of the eleventh ACM international workshop on data engineering for wireless and mobile access. ACM, pp 1–6
16. Shekhar S, Yang K, Gunturi VM, Manikonda L, Oliver D, Zhou X, George B, Kim S, Wolff JM, Lu Q (2012) Experiences with evacuation route planning algorithms. Int J Geog Inf Sci 26(12):2253–2265
17. Shi Q, Abdel-Aty M (2015) Big data applications in real-time traffic operation and safety monitoring and improvement on urban expressways. Transp Res Part C: Emerg Technol 58:380–394
18. Vlahogianni EI, Park BB, van Lint J (2015) Big data in transportation and traffic engineering. Transp Res Part C 58:161
19. Yang K, Evans MR, Gunturi VM, Kang JM, Shekhar S (2014) Lagrangian approaches to storage of spatio-temporal network datasets. IEEE Trans Knowl Data Eng 26(9):2222–2236

Chapter 2
Basic Graph Algorithms

2.1 A Brief Introduction to Graph Theory

A graph $G(N, E)$ consists of two finite sets N and E. The elements of N are called nodes and the elements of E are called edges of G. If an edge $e(a, b)$ joins nodes a and b, then a and b are incident (or adjacent) with edge $e(a, b)$. A directed edge is an edge where one incident node (or endpoint) is designated as the tail and the other incident node (or endpoint) is designated as the head. A directed graph (or digraph) is a graph each of whose edges is directed.

A line graph $L(G)$ of G is a graph where each node of $L(G)$ represents an edge of G and two nodes of $L(G)$ are adjacent if and only if they are adjacent as edges in G. A bipartite graph $G(N, E)$ is a graph whose node-set N can be partitioned into two subsets N_1 and N_2 such that each edge of G has one endpoint in N_1 and one endpoint in N_2. A path P_k is a non-empty graph $P_k(N, E)$ of the form $N = n_1, n_2, \ldots, n_k$ and $E = e(n_1, n_2), e(n_1, n_2), \ldots, e(n_{k-1}, n_k)$ where all nodes are all distinct. Two or more paths are independent if none of them contains the inner nodes ($n \in N - n_1, n_2$) of another. The distance $d(s, t)$ in G is the length of a shortest path from s to t.

A node-cut in a connected graph G is a node-set C whose removal disconnects the graph G. A cut-node (or articulation node) is a node-cut consisting of a single node. Similarly, an edge-cut in a connected graph G is an edge-set D whose removal disconnects the graph G. A cut-edge is an edge-cut consisting of a single edge.

2.2 Network Representations

There are many ways to represent a network structure. In this section, we present the most common and well-known data structures for network representation.

© Springer International Publishing AG 2017
K. Yang and S. Shekhar, *Spatial Network Big Databases*,
DOI 10.1007/978-3-319-56657-3_2

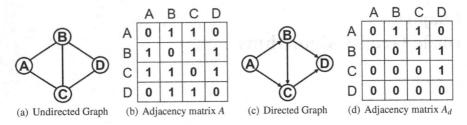

(a) Undirected Graph   (b) Adjacency matrix A (c) Directed Graph   (d) Adjacency matrix A_d

Fig. 2.1 Node-node adjacency matrix representation

2.2.1 Node-Node Adjacency Matrix

The node-node adjacency matrix A of an undirected graph G is a symmetric matrix whose rows and columns are defined as

$$A[i,j] = \begin{cases} 1 & \text{if node } i \text{ and node } j \text{ are adjacent} \\ 0 & \text{otherwise} \end{cases}$$

The node-node adjacency matrix can assign a weight (e.g., cost or capacity) of edges (i, j) on the ijth element in the matrix. Clearly, A is a symmetric matrix with zeros on the diagonal and only the upper (or lower)-triangular part of the matrix is stored. The sum of the elements in any ith row (or jth column) of the matrix is the degree of node i (or j). The power of A (i.e., A^k) is the number of walks of length k from i to j. Figure 2.1b shows an example of the node-node adjacency matrix (A) of the undirected graph in Fig. 2.1a.

The node-node adjacency matrix A_d of a directed graph G is a matrix whose rows and columns are defined as

$$A_d[i,j] = \begin{cases} 1 & \text{if there is a directed edge from node } i \text{ to node } j \\ 0 & \text{otherwise} \end{cases}$$

The sum of the elements of the ith row of the matrix is the out-degree of node i and the sum of the elements of the jth column of the matrix is the in-degree of node j. Figure 2.1d shows an example of the node-node adjacency matrix (A_d) of the directed graph in Fig. 2.1c. Let n be the number of nodes. The total space requirement of the node-node adjacency matrix is $O(n^2)$, which is inefficient if the graph is spare.

2.2.2 Node-Edge Incidence Matrix

The node-edge incidence matrix I of an undirected graph G is a matrix whose rows and columns are defined as

Fig. 2.2 Node-edge
incidence matrix
representation

	AB	AC	BC	BD	CD
A	1	1	0	0	0
B	1	0	1	1	0
C	0	1	1	0	1
D	0	0	0	1	1

(a) Incidence Matrix I

	AB	AC	BC	BD	CD
A	1	1	0	0	0
B	-1	0	1	1	0
C	0	-1	-1	0	1
D	0	0	0	-1	-1

(b) Incidence Matrix I_d

$$I[i,j] = \begin{cases} 1 & \text{if node } i \text{ is incident to edge } j \\ 0 & \text{otherwise} \end{cases}$$

The sum of the elements in the ith row of the matrix is the degree of node i and the sum of the elements in any column of the matrix is 2. Figure 2.2a shows an example of the node-edge incident matrix (I) of the undirected graph in Fig. 2.1a.

The incidence matrix I_d of an directed graph G is a matrix whose rows and columns are defined as

$$I_d[i,j] = \begin{cases} -1 & \text{if node } i \text{ is the tail of edge } j \\ 1 & \text{if node } i \text{ is the head of edge } j \\ 0 & \text{otherwise} \end{cases}$$

The node-edge incidence matrix $I_d[i, j]$ represents the orientation of a directed graph G. Figure 2.2b shows an example of the node-edge incidence matrix (I_d) of the directed graph in Fig. 2.1c. Let n be the number of nodes and let m be the number of edges. The total space requirement of the node-node adjacency matrix is $O(n \cdot m)$, which is also inefficient if the graph is spare.

2.2.3 Adjacency List

Both the incidence matrix and adjacency matrix are inefficient for sparse graphs because most of the elements in the matrix are zeros. To remedy this issue, the adjacency-list representation of a graph G uses a linked-list (or array) to represent a set of incident nodes (or edges) of the node. Figure 2.3 shows examples of the adjacency list representation corresponding to the examples in Fig. 2.1a, c.

The adjacent list is more space-efficient than the adjacency (or incidence) matrix. However, it takes linear time to test whether a node (or edge) is incident to node n whereas both the node-node adjacency matrix and node-edge incident matrix take constant time.

Fig. 2.3 Adjacency list
representation

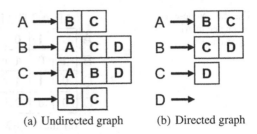

(a) Undirected graph   (b) Directed graph

Fig. 2.4 Forward star
representation

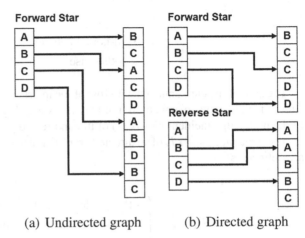

(a) Undirected graph   (b) Directed graph

2.2.4 Forward Star

The forward star representation is similar to the adjacency list representation. It uses
a singe array to store a set of incident nodes (or edges) and maintains a pointer for
each node that indicates the start address of a set of incidents. For a directed graph,
we construct the forward star to represent the outgoing edges for each node and the
reverse star to represent the incoming edges for each node. Figure 2.4 shows forward
star (or reverse star) representations corresponding to the examples in Fig. 2.1a, c.

The forward star representation is more space-efficient than the adjacent list representation. However, it is more difficult to update the graph structure whereas the
adjacency list representation requires constant time for addition and linear-time for
update and deletion.

2.3 Shortest Paths

This section reviews the computational methods for finding shortest paths between
nodes on a weighted graph. The shortest path computation is the most important component in developing spatial network queries including nearest neighbors, distance

estimation, the shortest route, etc. It also serves as a building block for developing more complex network queries. For example, the max-flow computation can be solved by iteratively finding a shortest path on the residual network (see Sect. 2.5).

2.3.1 Single-Source Shortest Path (SSSP)

Given a directed graph $G(N, E)$ where every edge $e \in E$ has a weight (or cost) $c(e)$, the Single-Source Shortest Path (SSSP) problem is to find a shortest simple path from a give source node s to every other nodes $n \in N$. If the graph includes a negative-weight cycle, the problem is NP-complete [12]. If no negative-weight cycle exists, there is a polynomial-time algorithm [2, 5].

The shortest path algorithms can be divided into two major groups: (1) Label-Correcting (LC) methods and (2) Label-Setting (LS) methods. Both approaches maintain a distance label $d(n \in N)$ for every node, which serves as an upper bound on the shortest path distance to node n. The Label-Correcting (LC) method can solve the shortest path problem in the general case in which the edge weights may be negative. It uses local information (i.e., incident nodes) of every node and updates (or reduces) the distance label (i.e., $d(n)$) of every node at each iteration. If no negative-weight cycle exists, the LC method finds the optimal solution in polynomial time. Examples of LC method include the Bellman-Ford-Moore algorithm [4, 11, 18]. Figure 2.5b shows an example of each step of the Bellman-Ford-Moore algorithm. Figure 2.5a illustrates the input with a transportation network (four nodes and five edges). Every edge is associated with a distance (e.g., travel time), as indicated by the number displayed alongside it. Assume that node A is a source node of G. First, the Bellman-Ford-Moore algorithm initializes the distance label of the source node ($d_1(A)$) to 0 and all others to ∞. Let the predecessor of n be n_{pre}. Then, the algorithm iteratively deceases $d_{i+1}(n)$ according to the following condition.

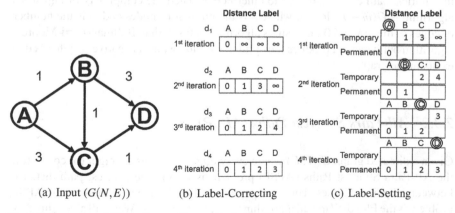

(a) Input ($G(N,E)$)   (b) Label-Correcting   (c) Label-Setting

Fig. 2.5 Examples of Single-Source Shortest Path algorithms

$$d_{i+1}(n) = min(d_i(n), d_i(n_{pre}) + c(n_{pre}, n)), \qquad (2.1)$$

where $c(n_{pre}, n)$ is the edge weight (or cost) of the edge from node n_{pre} to node n.

In the example in Fig. 2.5b, the second iteration updates the distance labels of all nodes as follows:

- $d_2(A) : min(d_1(A)) = 0$
- $d_2(B) : min(d_1(B), d_1(A) + c(AB)) = 1$
- $d_2(C) : min(d_1(C), d_1(A) + c(AC), d_1(B) + c(BC)) = 3$
- $d_2(D) : min(d_1(D), d_1(B) + c(BD), d_1(C) + c(CD)) = \infty$

This process continues until no further update is possible. The fourth iteration shows the shortest path distance from s to other nodes. The time complexity of the Bellman-Ford-Moore algorithm is $O(n \cdot m)$, where n is the number of nodes and m is the number of edges of the graph.

The Label-Setting (LS) method can solve the shortest path problem if the edge weights are non-negative. It maintains temporary and permanent distance labels on every node. The basic idea of the algorithm is to fan out from source node s and permanently label other nodes in the order of their distances from node s. The algorithm selects a node $n \in N$ with the minimum temporary label, makes it permanent, and scans out-going edges of node n to update (or reduce) the temporary distance labels of incident nodes. The algorithm terminates when all nodes are permanently labeled. The output of the LS method is a shortest-path tree. It grows a shortest-path tree, starting at node s, by adding an outgoing-edge whose endpoint is as close as possible to s at each iteration. Examples of LS methods include Dijkstra's algorithm and the A^* algorithm [8, 13]. Figure 2.5c shows an example of each step of Dijkstra's algorithm. First, Dijkstra's algorithm selects the source node A, permanently labels the distance of A, and updates the temporary distance labels of B and C. Then, it selects the node that has the minimum temporal distance label (i.e., B), permanently labels the distance of B, and update the temporary distance labels of C and D. This process continues until all nodes are permanently labeled. The forth iteration shows the shortest path distance from s to other nodes. The time complexity of Dijkstra's algorithm is $O(m + n \cdot \log n)$, where n is the number of nodes and m is the number of edges of the graph [20]. Dijkstra's algorithm is faster than Bellman-Ford-Moore's algorithm but it can not produce the optimal solution when a negative weighted edge exists in the graph.

2.3.2 All-Pairs Shortest Paths (APSP)

Given a directed graph $G(N, E)$ where every edge $e \in E$ has a weight (or cost) $c(e)$, the All-Pairs Shortest Paths (APSP) problem is to find the shortest path distance between all pairs of nodes. The most popular and well-known method for the APSP problem is the Floyd-Warshall algorithm [10, 23]. The Floyd-Warshall algorithm creates a node-node adjacency matrix and uses the concept of dynamic programming to find all-pairs shortest paths on a graph. It runs in $O(n^3)$ time, which is asymptotically no better than n calls to the SSSP algorithm from each node. For a sparse graph with non-negative edge weights, we could solve the all-pairs shortest paths by calling Dijkstra's algorithm from each node. This gives us an $O(m \cdot n + n^2 \cdot \log n)$ algorithm [20]. If the graph contains negative-weight edges, we can re-write the graph using the node potential function such that all edge weights are non-negative. The details about this function are present in Sect. 2.6. This leads us to apply Dijkstra's algorithm to the APSP problem with negative-weighted edges [14].

2.4 Block Decomposition

Spatial network queries for connectivity are important to measure the robustness and resilience of network structures. The vulnerable points (or cut points) in a network can be identified efficiently through a block decomposition technique. A graph $G = (N, E)$ is connected if any two nodes in G are linked by a path. A connected graph G is called bi-connected if for every node $n \in N$, $G - n$ is connected. An articulation node of a graph G is a node n such that $G - n$ is disconnected. A block of a graph is a maximal connected sub-graph H such that no node of H is an articulation node of H [7]. Consider the example in Fig. 2.6a. The removal of node D makes graph $G(N, E)$ disconnected. We can see that there are two articulation nodes (i.e., D and E). A block is a maximal connected subgraph without an articulation node. Graph G can be decomposed into these blocks which can show the overall structure of G using a tree.

Figure 2.6 shows examples of each step of the Depth-First-Search (DFS)-based algorithm [22]. First, it performs a Depth-First-Search (DFS), starting from an arbitrary node in G, and labeling nodes in the order in which they are discovered. We refer to this label as a *dnum*. Figure 2.6b shows the *dnum* for every node. The edges of the original graph can be divided into two types: tree edges and back edges. A tree edge belongs to the DFS-spanning tree itself; it connects a node to one of its descendants whereas a back edge connects a node to one of its ancestors. Let e be a back edge whose endpoints are a and b. Then $dnum(a) < dnum(b)$. Figure 2.6b shows tree edges (solid lines) and back edges (dotted lines) based on the original graph. Let tree T be the output of applying DFS to G. Then a non-root node n of T is

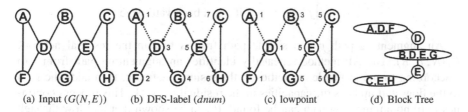

(a) Input ($G(N,E)$)   (b) DFS-label (*dnum*)   (c) lowpoint   (d) Block Tree

Fig. 2.6 Block decomposition

an articulation node of G if and only if node n has a child n_c such that no descendant of n_c has a path to an ancestor of node n by a back edge.

Let a set of tree edges be TE and let a set of back edges be BE. Let us define *lowpoint*(*a*) as following.

$$lowpoint(a) = min((lowpoint(b) \text{ where } ba \in TE) \cup (dnum(b) \text{ where } ab \in BE))$$

Figure 2.6c shows the *lowpoint* for every node. Given a tree edge ab, we can test whether node a is an articulation node as shown by the following condition: node a is an articulation node if and only if $dnum(a) \leq lowpoint(b)$ or node a is on at least two tree edges. Consider the example in Fig. 2.6. We can easily find two articulation nodes (i.e., D and E) because $dnum(D) < lowpoint(G)$ and $dnum(E) = lowpoint(H)$. Figure 2.6d shows a block decomposition containing three blocks.

2.5 Maximum Network Flow

Given a graph $G(N, E)$, a set of capacity constraints on edges $c(e \in E) \in C$, a source node s, and a sink node t, the maximum network flow problem is to find a flow of maximum value that meets capacity constraints. This general problem arises in many real applications (e.g., network flow analysis) and efficient algorithms for computing the maximum flow are critical for handling large-sized spatial networks. This section presents two general methods for solving the maximum-flow problem: (1) the Augmenting-Path (AP) method and (2) the Push-Relabel (PR) method.

2.5.1 Augmenting-Path Algorithm

The Augmenting-Path (AP) method is based on the idea of a residual network and augmenting paths [2]. A residual network represents the available capacities ($r \in R$) on the current network flow f. Let $c(e)$ be a capacity on edge e and let $f(e)$ be a flow on edge e. The residual capacity $r(e)$ on edge e can be defined as follows

$$r(e) = \begin{cases} c(e) - f(e) & \text{if } e \text{ is a forward edge} \\ f(e) & \text{if } e \text{ is a backward edge} \end{cases}$$

An augmenting path p is a simple path from s to t in the residual network $G_r(N, E, R)$. The AP method repeatedly identifies an augmenting path from s to t according to the available capacities in the residual network, then adds the path to the flow, and updates the capacities in the residual network. This process continues until no augmenting path can be found. It can be shown that the flow through

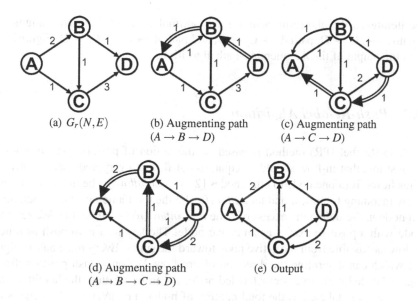

(a) $G_r(N,E)$ (b) Augmenting path   (c) Augmenting path
                      $(A \rightarrow B \rightarrow D)$ $(A \rightarrow C \rightarrow D)$

(d) Augmenting path   (e) Output
$(A \rightarrow B \rightarrow C \rightarrow D)$

Fig. 2.7 Augmenting Path algorithm

a network is optimal if and only if it contains no augmenting path on the residual network [2].

Let f be a flow in G. A feasible (or admissible) flow f in G satisfies the following two conditions:

1. Capacity Constraint: $f(e) \leq c(e)$, for every edge e in G.
2. Conservation Constraint: $\sum_{e \in In(n)} f(e) = \sum_{e \in Out(n)} f(e)$ for every node $n \in N - s, t$, where $In(n)$ is a set of incoming edges of node n and $Out(n)$ is a set of outgoing edges of node n.

Assume that N is divided into two sets, $s \in N_1$ and $t \in N_2$. Let the pair (N_1, N_2) be a cut in G and $c(N_1, N_2)$ be the capacity of this cut. It is well known that the maximum total value of a flow equals the minimum capacity of a cut [11].

Consider the example of AP method in Fig. 2.7. Figure 2.7a illustrates the residual network where every edge is associated with a capacity, as indicated by the number displayed alongside it. Given the residual network, the algorithm identifies the augmenting path on the available capacities and updates the residual network. We define the residual capacity of the augmenting path as the minimum residual capacity of any edge in the path. In each iteration, the algorithm finds an augmenting path p and updates the flow on each edge of p by the residual capacity of the augmenting path. In this example, the first iteration identifies path $A \rightarrow B \rightarrow D$ as an augmenting path and computes the residual capacity of the path (i.e., 1) (see Fig. 2.7b). This augmentation reduces the capacity of edge AB by 1 and increases the capacity of edge BA by 1. It also reduces the capacity of edge BD by 1 and increases the capacity of edge DB by 1. Figure 2.7b shows the residual network after augmenting path $A \rightarrow B \rightarrow D$. After three iterations, the algorithm achieves the optimal solution (Fig. 2.7e), augmenting path $A \rightarrow B \rightarrow D$, path $A \rightarrow C \rightarrow D$, and path $A \rightarrow B \rightarrow C \rightarrow D$. Figure 2.7e shows the output of the Augmenting-Path algorithm.

2.5.2   Push-Relabel Algorithm

The Push-Relabel (PR) method is based on the notion of pre-flow which ignores the constraint that in-flow should be equal to out-flow at every node and iteratively pushes flows from one node to other nodes [2, 5]. Let $inflow(n)$ be the total amount of flow incoming to node n and let $outflow(n)$ be the total amount of flow outgoing from node n. We define the excess of node n as $inflow(n) - outflow(n)$. We refer to a node with a positive excess as an active node. The key idea is to push as much pre-flow as possible from the active node toward the sink t. Every node has a height label which can determine the direction of pre-flow. We only push pre-flow from a higher labeled node to a lower labeled node. As an initial step, the height of the source (i.e., s) is labeled as the total number of nodes (i.e., $|N|$) and the heights of others are labeled as 0. The height of a node can be increased to one unit more than the height of the lowest of its incident nodes when the node cannot push all excess flow to its incident nodes. The algorithm selects an active node, pushes pre-flow from the active node until either the node's excess becomes zero or the algorithm relabels the node. The algorithm terminates when the network contains no active node.

Consider the example in Fig. 2.8. Figure 2.8a illustrates the residual network where every edge is associated with a capacity, as indicated by the number displayed alongside it. Given the residual network, the algorithm labels the heights of every node (see Fig. 2.8b). Then, it chooses node A as an active node and pushes excess flow to incident nodes (i.e., B and C) (see Fig. 2.8c). Figure 2.8d shows the residual network after pushing excess flow from node A.

Next, the algorithm chooses node B as an active node. Since no outgoing-edge exists for pre-flow, the height of node B increases by 1 (see Fig. 2.9b). Then, it pushes excess flow from node B to incident nodes (i.e., C and D) (see Fig. 2.9b). Figure 2.8d shows the residual network after pushing excess flow from node B.

After that, the algorithm chooses node C as an active node, increases the height by 1, and pushes the excess flow to node D. Since there is no excess node except the

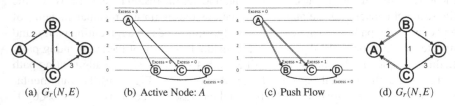

(a) $G_r(N, E)$ (b) Active Node: A (c) Push Flow   (d) $G_r(N, E)$

Fig. 2.8 Push Relabel algorithm: 1st iteration

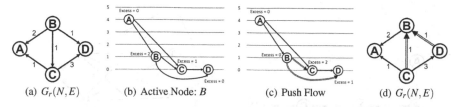

Fig. 2.9 Push Relabel algorithm: 2nd iteration

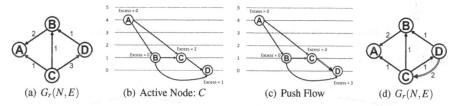

Fig. 2.10 Push Relabel algorithm: 3rd iteration

sink t, the algorithm terminates. Figure 2.10d shows the output of the Push-Relabel algorithm.

2.6   Bipartite Weighted Matching

Graph $G(N, E)$ is bipartite if the nodes can be divided into two sets, G_1 and G_2, such that all edges ($e \in E$) have one node in G_1 and one node in G_2. A matching in a graph $G(N, E)$ is a subset of edges $\hat{E} \subset E$ such that no two edges of $\hat{E}$ share a node. Given a weighted bipartite network $G = (N_1 \cup N_2, E)$ with $|N_1| = |N_2|$ and edge weights $w(e)$, the Bipartite Weighted Matching problem is to find the perfect matching with the minimum total weight. The problem is important in many practical applications, such as resource assignment. In this section, we introduce the Successive Shortest Path algorithm that can produce the optimal solution for the Bipartite Weighted Matching problem [2].

Consider the resource assignment problem in Fig. 2.11. Figure 2.11a shows an example input network consisting of a graph with 4 graph-nodes (A, B, C, D) and two service center nodes (X, Y) with capacities of 2 each. The objective of this problem is to assign graph-nodes to a service center that minimizes the total distance between graph-nodes (N_1) to service center nodes (N_2). We assume that every service center node can be matched with exactly two graph-nodes. Figure 2.11b shows a bipartite graph, consisting of two sets: graph nodes (N_1) and service center nodes (N_2). For efficiency, we create a super-source node S that is connected to every node in N_1 by an edge of weight 0 and create a sink node T and connect it to every node in N_2 by an edge of weight 0. Every edge is associated with the distance from a graph-node to a service center node, as indicated by the number displayed alongside it.

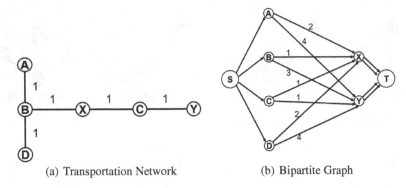

| (a) Transportation Network | (b) Bipartite Graph |

Fig. 2.11 Example of resource assignment

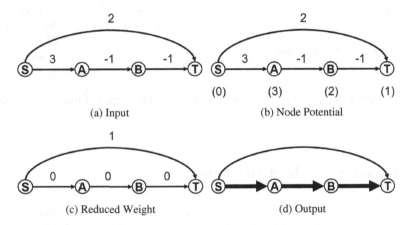

| (a) Input | (b) Node Potential |

| (c) Reduced Weight | (d) Output |

Fig. 2.12 Node potential and reduced weight

The matching problem in Fig. 2.11b is identical to the Bipartite Weighted Matching problem if each service center node can be split into two nodes. In this example, we use the Successive Shortest Path algorithm to find the optimal matching.

The Successive Shortest Path algorithm uses three important notions: node potential, reduced weight function, and shortest augmenting path. Given a weighted, directed graph $G(N, E)$ with edge weights $w(e)$, let $p(a)$ be a shortest path distance from source node s to node a. We refer to $p(a)$ as the node potential of node a. Then the reduced weight function can be defined as follows: $w(a, b) = w(a, b) + p(a) - p(b)$. Consider the example in Fig. 2.12a. The shortest path from S to T is $S \rightarrow A \rightarrow B \rightarrow T$. We can use the Label-Correcting (LC) method to compute the shortest path in the graph with negative weighted edges, but the Label-Setting (LS) method usually outperforms the Label-Correcting (LC) method. If the reduced weight function converts all edge-weights into non-negative edge-weights, more efficient algorithms (i.e., the LS method) can be applied to find the shortest path. Figure 2.12b shows the node potential for every node and Fig. 2.12c shows the reduced weight on every edge.

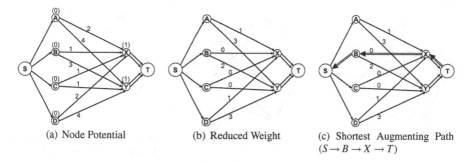

(a) Node Potential   (b) Reduced Weight   (c) Shortest Augmenting Path
$(S \rightarrow B \rightarrow X \rightarrow T)$

Fig. 2.13 Successive Shortest Path algorithm: 1st iteration

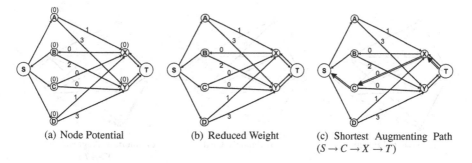

(a) Node Potential   (b) Reduced Weight   (c) Shortest Augmenting Path
$(S \rightarrow C \rightarrow X \rightarrow T)$

Fig. 2.14 Successive Shortest Path algorithm: 2nd iteration

Since all edge-weights are non-negative, we can easily find the shortest path using the LS method. A node potential is a generalization of the concept of distance label that we used in the Successive Shortest Path algorithm.

Consider the example in Fig. 2.11b. We assume that every edge has a unit capacity. Figure 2.13a, b show the node potentials and reduced weights produced from Fig. 2.11b. In this example, we augment the shortest path $S \rightarrow B \rightarrow X \rightarrow T$ and update the residual network (see Fig. 2.13c).

Next, we update note potentials (see Fig. 2.14a) and reduce the edge-weights (see Fig. 2.14b). Then we augment the shortest path $S \rightarrow C \rightarrow X \rightarrow T$ and update the residual network (see Fig. 2.14c). After two more iterations (see Figs. 2.15 and 2.16), the Successive Shortest Path algorithm produces the optimal solution for the Bipartite Weighted Matching problem (i.e., nodes A and B are assigned to node X and nodes C and D are assigned to node Y). It is well known that the Bipartite Weighted Matching problem is a special case of the minimum cost network flow problem [2].

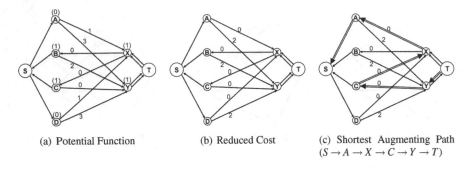

(a) Potential Function   (b) Reduced Cost   (c) Shortest Augmenting Path
                                                          $(S \rightarrow A \rightarrow X \rightarrow C \rightarrow Y \rightarrow T)$

Fig. 2.15 Successive Shortest Path algorithm: 3rd iteration

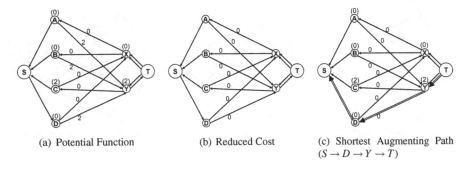

(a) Potential Function   (b) Reduced Cost   (c) Shortest Augmenting Path
                                                          $(S \rightarrow D \rightarrow Y \rightarrow T)$

Fig. 2.16 Successive Shortest Path algorithm: 4th iteration

2.7 Graph Partitioning

Recently, big data processing platforms (e.g., MapReduce, Apache Hadoop, Apache Spark and GraphLab) have become a growing component for large-scale data management due to their potential ability to distribute large datasets across multiple machines and parallelize the query processing [1]. Data decomposition for parallel query processing follows three general rules to process the query efficiently: (1) data should be placed as close as possible to where it will be used, (2) load balancing should be considered to increase the global system performance, and (3) low communication cost should be considered to improve the query response. Graph Partitioning is a critical step to meet these general rules. For example, big data processing platforms can reduce computational cost for data processing when large sized datasets can be divided into smaller ones of about equal size, with minimum interaction as possible between these smaller datasets. Graph Partitioning decomposes large-sized networks into non-overlapping sub-networks which can efficiently apply a divide-and-conquer scheme to spatial network query optimization (e.g., shortest path computation) [6, 15, 21]. Given a spatial network and a number of partitions (i.e., k), the Graph Partitioning problem is to decompose the spatial network into k non-overlapping equal-size sub-networks that minimizes the total weight of the edge-cuts between partitions. The problem is NP-hard for the case k = 2, which is also called the Minimum Bisection problem. In this section, we review the most popular methods to solve the Graph Partitioning problem.

The basic approach for dealing with Graph Partitioning is to construct an initial solution and iteratively improve the solution with a local search [3]. The Spectral method has been widely used to produce a good initial solution [19]. The Spectral method finds an approximate solution to the Graph Partitioning problem by computing the eigenvectors of a Laplacian Matrix and inferring the edge-cuts from these eigenvectors. The Laplacian Matrix of a graph G is defined as $L = D - A$, where D is the diagonal matrix expressing node degree and A is the node-node adjacency matrix.

$$L[i,j] = \begin{cases} degree(i) & \text{if } i = j \\ -1 & \text{if } i \neq j \text{ and there is an } edge(i,j) \\ 0 & \text{otherwise} \end{cases}$$

The general idea of the Spectral method is to partition the network by utilizing the eigenvector v_2 corresponding to the second smallest eigenvalue λ_2 of the Laplacian matrix. Consider the example for the Minimum Bisection problem in Fig. 2.17a. Figure 2.17b shows the Laplacian Matrix for the given input graph. The second smallest eigenvalues λ_2 is 0.191 and its corresponding eigenvector v_2 is shown in Fig. 2.17c. The main assumption for the Spectral method is that the element of the eigenvector v_2 is binary (i.e. 1 or -1) to represent each partition. In general, elements are not binary, but are distributed over a range of real values, as shown in Fig. 2.17c. One of the ways to handle this issue is to find the median value of elements in the eigenvector and use it as the threshold to bisect a set of nodes. In this example, the median value is 0 and nodes with positive value become one partition and nodes with negative value become the other. Since node B and G have the median value (i.e., 0), these nodes are arbitrarily allocated to one of the two partitions. Figure 2.17d shows the output of the Spectral method.

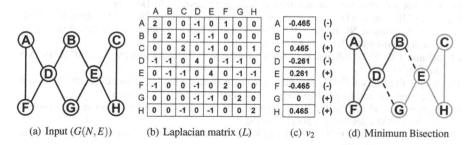

| (a) Input ($G(N,E)$) | (b) Laplacian matrix (L) | (c) v_2 | (d) Minimum Bisection |

Fig. 2.17 Spectral graph partitioning method

The Kernighan-Lin (KL) and Fiduccia-Mattheyses (FM) methods have been widely used to iteratively improve the solution quality by local search optimization [9, 17]. These methods exchange pairs of nodes, each node from a different partition, and reduce the edge-cuts between partitions. If the network size is too large, Multilevel Partitioning techniques are used to reduce the size of the input network [16]. The Multilevel approach consists of three main phases: (1) coarsening, (2) initial partitioning, and (3) uncoarsening. In the coarsening phase, it iteratively identifies matchings and contracts the edges to reduce the size of the input. For example, Graph G_{i+1} is constructed from G_i by finding a maximal matching of G_i and aggregating the nodes that are incident on each edge of the matching. After that, it constructs an initial partition using spectral or KL/FM methods. In the uncoarsening phase, the partition is projected back to the original graph by progressively disaggregating the nodes. Local refinement can be applied in each step of uncoarsening to improve the solution quality.

References

1. Agneeswaran VS (2014) Big data analytics beyond hadoop: real-time applications with storm, spark, and more hadoop alternatives. FT Press, Upper Saddle River
2. Ahuja R, et al (1993) Network flows: theory, algorithms, and applications. Prentice Hall, Upper Saddle River
3. Alpert C, Kahng A (1995) Recent directions in netlist partitioning: a survey. Integr VLSI J 19(1–2):1–81
4. Bellman R (1958) On a routing problem. Q Appl Math 16:87–90
5. Cormen TH (2009) Introduction to algorithms. MIT Press, Cambridge
6. Demetrescu C, Goldberg AV, Johnson DS (2009) The shortest path problem: ninth DIMACS implementation challenge, vol 74. American Mathematical Society, Providence
7. Diestel R (2005) Graph Theory, (4th edn). Graduate texts in mathematics, vol 173. Springer, Heidelberg
8. Dijkstra EW (1959) A note on two problems in connexion with graphs. Numerische mathematik 1(1):269–271
9. Fiduccia C, Mattheyses R (1982) A linear-time heuristic for improving network partitions. In: 19th conference on design automation. IEEE, pp 175–181
10. Floyd RW (1962) Algorithm 97: shortest path. Commun ACM 5(6):345
11. Ford LR Jr (1956) Network flow theory. Technical report, DTIC Document
12. Garey MR, Johnson DS (2002) Computers and intractability, vol 29. WH Freeman, New York
13. Hart PE, Nilsson NJ, Raphael B (1968) A formal basis for the heuristic determination of minimum cost paths. IEEE Trans Syst Sci Cybern 4(2):100–107
14. Johnson DB (1977) Efficient algorithms for shortest paths in sparse networks. J ACM (JACM) 24(1):1–13
15. Jung S, Pramanik S (2002) An efficient path computation model for hierarchically structured topographical road maps. IEEE Trans Knowl Data Eng 14(5):1029–1046
16. Karypis G et al (1998) A fast and high quality multilevel scheme for partitioning irregular graphs. SIAM J Sci Comput 20(1):359–392
17. Kernighan B, Lin S (1970) An efficient heuristic procedure for partitioning graphs. Bell Syst Tech J 49(2):291–307
18. Moore EF (1959) The shortest path through a maze. Bell Telephone System, New York
19. Nascimento MC, De Carvalho AC (2011) Spectral methods for graph clustering-a survey. Eur J Oper Res 211(2):221–231

20. Schrijver A (2002) Combinatorial optimization: polyhedra and efficiency, vol 24. Springer Science & Business Media, Heidelberg
21. Shekhar S, Fetterer A, Goyal B (1997) Materialization trade-offs in hierarchical shortest path algorithms. In: International symposium on spatial databases. Springer, pp 94–111
22. Tarjan R (1971) Depth-first search and linear graph algorithms. In: 12th annual symposium on switching and automata theory, pp 114–121. doi:10.1109/SWAT.1971.10
23. Warshall S (1962) A theorem on boolean matrices. J ACM (JACM) 9(1):11–12

Chapter 3
Capacity Constrained Network Voronoi Diagrams

3.1 Introduction

Given a graph and a set of service center nodes (e.g., gas stations) with capacity constraints (e.g., amount of gasoline, size of parking lot, etc.), a Capacity Constrained Network-Voronoi Diagram (CCNVD) partitions the graph into a set of contiguous service areas (SAs) that honor service center capacities and minimize the sum of the shortest distances from graph-nodes to allotted service centers. Figure 3.1a shows an example input of CCNVD consisting of a graph with 15 graph-nodes ($A, B, \ldots, O$) and three service center nodes (X, Y, and Z) with capacities of 5 each. Figure 3.1b shows an example output of CCNVD where the graph is partitioned such that 5 graph-nodes are allotted to each service center, as shown by the distinct colors. In essence, the CCNVD problem is to partition a spatial network into non-overlapping sub-networks which can assign users to service centers while meeting contiguity and capacity constraints and minimizing the total shortest distance from users to assigned facilities. The CCNVD problem is NP-hard and exhibits NPO-completeness (a proof is provided in Sect. 3.1.3). Intuitively, the problem is computationally challenging because of the large size of the transportation network and the contiguity constraint.

3.1.1 Application Domains

The CCNVD problem is important for critical applications such as assigning people to relief-supply distribution centers in the aftermath of a disaster, assigning evacuees to shelter facilities, and assigning jurisdictions to emergency responders such as hospitals and police stations. CCNVD provides a contiguous, compact, and simple representation of service areas that mitigates resource shortages and allow for efficient delivery of critical information to the public. CCNVD may also be used for shelter allocation where contiguous service areas reduce movement conflicts (which raise the risk of congestion, stampedes, etc.) across people heading to different shel-

© Springer International Publishing AG 2017
K. Yang and S. Shekhar, *Spatial Network Big Databases*,
DOI 10.1007/978-3-319-56657-3_3

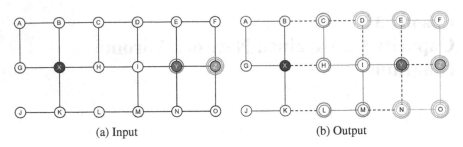

(a) Input   (b) Output

Fig. 3.1 Example of the Input and Output of CCNVD (*Colors* show service center allotment) (Color figure online)

ters. By providing contiguous service areas, it also promotes clear communication of official emergency instructions.

3.1.2 Problem Definition

In the formulation of the CCNVD problem, a transportation network is represented and analyzed as an undirected graph composed of nodes and edges. Each node represents a spatial location in geographic space (e.g., road intersections), which can be used as a proxy for locations of citizens or residences. Each edge between two nodes represents a road segment and has a travel distance. Each service center has a given capacity (the number of people it can efficiently serve). The $CCNVD(N, E, S, C, D)$ problem is defined as follows:

Input: A transportation network G with

- a set of graph-nodes N and a set of edges E,
- a set of fixed service center locations $S \subset N$,
- a set of positive integer capacities for service centers $C : S \rightarrow \mathbb{Z}^+$, and
- a set of nonnegative real distances of edges $D : E \rightarrow \mathbb{R}_0^+$

Output: A Capacity Constrained Network Voronoi Diagram (CCNVD)
Objective:

- Min-sum: Minimize the sum of the shortest distances from graph-nodes to their allotted service centers.

Constraints:

- Capacity Constraint: Each service area (SA) contains exactly one service center and the number of graph-nodes in the SA does not exceed the capacity of the SA.
- Contiguity Constraint: Each service area (SA) should be a connected sub-graph of G.

3.1.3 Problem Hardness

The CCNVD problem is related to the connected k-partition problem, which is known to be NP-hard [8]. In addition, the CCNVD problem has NPO-completeness and thus no approximation is possible in polynomial time, unless $P = NP$. The proof of hardness is provided in the following.

Definition 1 (**Connected k-partition problem**) Given a graph $G = (N, E)$, service centers $s_1, s_2, \ldots, s_k \in N$, and positive integer capacities for service centers $c_1, c_2, \ldots, c_k$, where $c_1 + c_2 + \ldots + c_k = |N|$, the connected-$k$-partition $(k-CP(N, E, S, C))$ problem separates the service centers $s_1, s_2, \ldots, s_k$, and partition p_i containing s_i is a connected sub-graph consisting of c_i nodes for $i = 1, 2, \ldots, k$ (e.g., equal sized connected sub-graphs). It has been proved that a connected-k-partition of N is a NP-hard problem [8].

Solution Existence: It is also known that a solution exists for k-connected graphs, formally stated as follows:

Let $G = (N, E)$ be a k-node-connected graph. Let $s_1, s_2, \ldots, s_k \in N$, $c = |N|$, and let $c_1, c_2, \ldots, c_k$ be positive integers such that $c_1 + c_2 + \ldots + c_k = c$. Then there exists a k-partition of $N(G)$ such that this partition separates the nodes $s_1, s_2, \ldots, s_k$, and the partition p_i containing s_i is a connected sub-graph consisting of c_i vertices, for $i = 1, 2, \ldots, k$ [13, 14, 17].

Theorem 1 *The CCNVD problem is NP-hard [20].*

Proof The CCNVD problem belongs to NP since, given an instance of a CCNVD and a maximum bound T, we can take a set of connected sub-graphs such that the sum of the shortest distances from nodes (N) to their allotted service centers (S) is lower than T as a valid certification. Let $A = (N, E, S, C)$ be an instance of a connected-k-partition problem, where N is a set of nodes, E is a set of edges, S is a set of service centers, and C is a set of capacities for service centers. Let $B = (N, E, S, C, D, T)$ be an instance of the CCNVD problem, where D is a set of distances of E, and T is a maximum bound of the sum of the shortest distances from nodes (N) to their allotted service centers (S). Then it is easy to show that a connected k-partition is a special case of CCNVD, where k is the number of service centers S, $d \in D$ has a distance value of zero, and T is unbounded. Since A is constructed from B in polynomial time, the proof is complete.

The NPO-completeness of CCNVD follows from a well-known result about the NP-hardness of the connected k-partition problem [3, 8], which partitions a graph into k connected sub-graphs where $k \geq 3$ (Fig. 3.2).

Theorem 2 *No polynomial-time approximation algorithm exists for CCNVD if $P \neq NP$ [21].*

Proof Assume P $\neq$ NP. Let $n = |N|$ and $k = |S|$. We construct a mapping from an instance $k-CP(N, E, S, C)$ of the connected k-partition problem to an instance

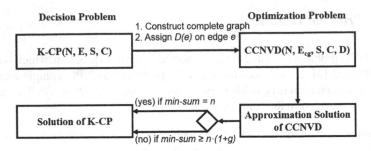

Fig. 3.2 Approximation hardness of CCNVD

$CCNVD(N, E, S, C, D)$ of the CCNVD problem, such that the question of whether $k-CP$ has a solution can be determined from any $CCNVD$ whose min-sum is n. Given a $k-CP(N, E, S, C)$, we add additional edges to $k-CP$ and construct a complete graph $CG(N, E_{cg}, S, C)$. Then the instance of $CCNVD(N, E_{cg}, S, C, D)$ can be constructed by assigning $D(e)$ on every $e(v_i, v_j) \in E_{cg}$ as follows:

$$D(e) = \begin{cases} 1 & \text{if } e \in E \text{ and } (v_i \in S \text{ or } v_j \in S) \\ 1 + n \cdot g & \text{if } e \notin E \\ 0 & \text{otherwise} \end{cases}$$

If the min-sum of $CCNVD$ is n, then the solution of $k-CP(N, E, S, C)$ exists because all nodes in $SA(s \in S)$ form a connected component. If the min-sum of $CCNVD$ is at least $n \cdot (1 + g)$, then no solution of $k-CP$ exists because at least one node is disconnected from $SA(s \in S)$. If there exists a polynomial time g approximation algorithm for $CCNVD$, then it can solve the NP-hard problem $k-CP$ in polynomial time, implying P=NP. This contradicts the original assumption that P $\neq$ NP, so no polynomial-time approximation exists for CCNVD.

3.1.4   Literature Review

Related work on minimizing the sum of the distances between graph-nodes and their allotted service centers can be categorized into two groups: (1) methods that ensure service area (SA) contiguity and (2) methods that honor service center capacity constraints. Related work on SA contiguity includes the creation of Network Voronoi Diagrams (NVDs), in which each node is assigned to its nearest service center by definition [9, 18]. Network Voronoi Diagram (NVD) is a special case of the CCNVD problem where the service center capacity is unlimited. However, NVDs were not designed to account for capacity constraints. Related work on honoring service center capacity constraints includes min-cost flow approaches [1, 6, 16]. However, these approaches do not always preserve service area contiguity. By contrast, Capacity

Fig. 3.3 Approaches to minimizing the sum of the shortest distances between nodes and their allotted service centers

		Service center capacity constraints honored	
		no	yes
Service Area Contiguity	no		Min-Cost Flow
	yes	Network Voronoi Diagram (nearest center)	Proposed Work

Constrained Network Voronoi Diagrams (CCNVD) create Network Voronoi Diagram (NVD) that honors both capacity and contiguity constraints (Fig. 3.3).

Figure 3.4 illustrates examples of related work. Figure 3.4a shows the input with a transportation network (15 graph-nodes $(A, B, \ldots, O)$ and three service centers $(X, Y, \text{and } Z)$). Every edge is associated with a distance (e.g., travel time), as indicated by the number displayed above it. For simplicity, every graph-node has one unit of demand. Every service center has a capacity to serve 5 units of graph-nodes. Figure 3.4b shows a Network Voronoi Diagram (NVD), allotting every graph-node to the nearest (e.g., shortest path) service center. NVD assigns 8 graph-nodes (color=blue) to service center X, 5 graph-nodes (color=red) to service center Y, and 2 graph-nodes (color=green) to service center Z. The dotted lines represent the boundary edge between two adjacent SAs. Although NVD allots every graph-node to the nearest service center, it does not account for load balance, which may lead to congestion and delay at service center X while service center Z may have few graph-nodes. Figure 3.4c shows one possible example of the min-cost flow approach that minimizes the sum of the distances from graph-nodes to their allotted service centers. Although it achieves its goal (min-sum), service areas for Y and Z violate SA contiguity. Figure 3.4d shows an example of the proposed Capacity Constrained Network Voronoi Diagram (CCNVD). As can be seen, the load is balanced: 5 nodes are allotted to every service center as shown by distinct colors, and service areas are contiguous for all service centers.

3.1.5 Outline of the Chapter

The rest of the chapter is organized as follows: Sect. 3.2 describes three algorithms and techniques that can reduce the computational costs for creating a CCNVD. Section 3.3 presents a case study using a Brooklyn, NY road map. Section 3.4 summarizes the chapter.

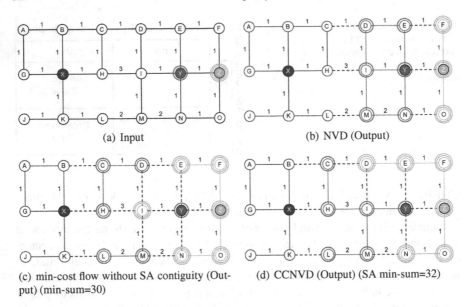

(a) Input

(b) NVD (Output)

(c) min-cost flow without SA contiguity (Output) (min-sum=30)

(d) CCNVD (Output) (SA min-sum=32)

Fig. 3.4 Example of the Input and Output of NVD, min-cost flow, and CCNVD (*Colors* show service center allotment) (Color figure online)

3.2 Algorithms for Capacity Constrained Network Voronoi Diagram

In this section, we describe three approaches to the CCNVD problem.

3.2.1 Pressure Equalizer (PE) Algorithm

The Pressure Equalizer (PE) algorithm starts with a Network Voronoi Diagram (NVD) and iteratively re-assigns graph-nodes until the capacity constraint is met. The first core idea in PE is to use an NVD as the initial iteration because (1) an NVD represents the optimal sum of the shortest distances under no capacity constraint and (2) an NVD creates contiguous service areas [20]. Thus, by starting with an NVD and keeping changes (re-assignments) to the NVD as minimal as possible, the PE algorithm can keep the sum of the shortest distances relatively low and preserve contiguity. The second core idea in PE is the Pressure Equalization Graph (PE-Graph), where pressure for a service center s refers to the difference between the capacity of s and the number of nodes allotted to s. Positive values of pressure indicate overload and negative values indicate slack or available capacity. The PE-nodes of a PE-Graph are service centers S in the transportation network. The PE-nodes are of three types: excess, deficit, and balanced. Let *capacity*(s) be the capacity of

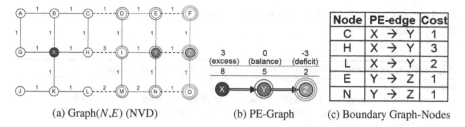

Node	PE-edge	Cost
C	X → Y	1
H	X → Y	3
L	X → Y	2
E	Y → Z	1
N	Y → Z	1

(a) Graph(N,E) (NVD)　　　　(b) PE-Graph　　　(c) Boundary Graph-Nodes

Fig. 3.5 PE: Iteration 1 (*Colors* show service center allotment) (Color figure online)

a PE-node $s \in S$ and *allotment*(s) be the number of graph-nodes allotted to s. If *allotment*(s) > *capacity*(s), we refer to s as an excess PE-node whose excess value is *allotment*(s) − *capacity*(s). On the other hand, if *allotment*(s) < *capacity*(s), we refer to s as a deficit PE-node whose deficit value is *capacity*(s) − *allotment*(s). We refer to a PE-node s with *allotment*(s) = *capacity*(s) as balanced. The collection of graph-nodes allotted to a PE-node s represents the service area (SA) for s. We refer to $SA(s)$ as the service area for s. A PE-edge is inserted from PE-node $s1$ to PE-node $s2$ if any allotted graph-node $n1$ on $s1$ (i.e., $n1 \in SA(s1)$) is connected to any allotted graph-node $n2$ on $s2$ (i.e., $n2 \in SA(s2)$). We refer to both $n1$ and $n2$ as boundary graph-nodes. Figure 3.5b shows the PE-Graph for the NVD of Fig. 3.5a. The graph has three PE-nodes for service centers X, Y, and Z and two PE-edges (i.e., $X \rightarrow Y$ and $Y \rightarrow Z$). PE-node X has an excess of 3 and PE-node Z has a deficit of 3.

The PE algorithm tries to satisfy capacity constraints for every service center, maintain service area contiguity constraints, and reduce the sum of the shortest distances from graph-nodes to their allotted service centers (PE-nodes). At each step, the algorithm re-allots a graph-node from an excess PE-node to fulfill the capacity constraint. The effect of re-allotting a graph-node n from $s1$ to $s2$ on an objective function (SA min-sum) can be defined using the following cost function:

$$Cost(a) = \sum_{\substack{s1 \rightarrow s2 \\ n \in SA(s1)-\{a\}}} sd(n, s1) - \sum_{n \in SA(s1)} sd(n, s1)$$

$$+ \sum_{n \in SA(s2)+\{a\}} sd(n, s2) - \sum_{n \in SA(s2)} sd(n, s2), \qquad (3.1)$$

where $sd(n, s)$ is the length of the SA shortest distance from n to s in $SA(s)$ [20].

A key idea behind PE is to first choose the best boundary graph-node that minimizes the cost of re-allotment and then re-allot this graph-node to fulfill capacity constraints. Figure 3.5c shows boundary nodes for the NVD of Fig. 3.5a. In this example, the best boundary nodes in terms of minimizing the re-allotment cost are node C for $X \rightarrow Y$ and node E (or N) for $Y \rightarrow Z$. There is one PE-path ($X \rightarrow Y \rightarrow Z$) to traverse from the excess PE-node X to the deficit PE-node Z. Thus, the algorithm re-allots node C from X to Y and node E from Y to Z to reduce *excess*(X) by 1 in the first iteration.

Algorithm 1 Generalized Pressure Equalizer (PE) Algorithm (Pseudo-code)

Inputs:
  - A transportation network ($Graph(N, E)$) with a set of graph-nodes N and edges E.
  - A set of PE-nodes (service centers) $S \subset N$ with their capacity C
  - Every edge has a distance $d(e)$
Outputs: Capacity Constrained Network Voronoi Diagram ($CCNVD$)
Steps:
1: Create an **initial partition** that preserves service area (SA) contiguity.
2: **while** Any PE-node $s \in S$ has excess graph-nodes **do**
3:   Create $PE-Graph(S, E_{pe})$ where PE-edge $e_{pe} \in E_{pe}$ connects two adjacent SAs.
4:   Find all boundary graph-nodes $N_{bdy} \subset N$ and compute re-allotment cost ($Cost_{s1 \rightarrow s2}(n_{bdy} \in N_{bdy})$).
5:   Find the best boundary graph-nodes $N_{best\_bdy} \subset N_{bdy}$ which minimize the re-allotment cost.
6:   Group all excess PE-nodes $S_{ex} \subset S$ with a super-source node src_{ex} and group all deficit PE-nodes $S_{df} \subset S$ with a super-sink node $sink_{df}$.
7:   Find the best PE-path p in terms of **preserving SA contiguity (i.e., SACC)** as well as minimizing the sum of re-allotment costs from src_{ex} to $sink_{df}$. If no PE-path is founded, then return "no solution found".
8:   Re-allot the best boundary graph-nodes (n_{best_bdy}) on the best path p.
9: **end while**
10: return CCNVD. i.e., final allotment of graph-nodes to their service centers.

Algorithm 1 presents the pseudo-code for a generalized version of PE. First, PE creates an initial partition that preserves service area (SA) contiguity (line 1). In this step, PE initializes CCNVD with NVD. It then creates a PE-Graph and finds all boundary graph nodes, as well as the best boundary graph nodes (lines 3–5). After that, it groups excess PE-nodes into a super-source node and groups deficit PE-nodes into a super-sink node (line 6). Next it searches the PE-graph and finds the best PE-path (line 7). The re-allotments through the best PE-path should preserve SA contiguity. Therefore, the PE-path computation part contains Service Area Contiguity Checking (SACC) (e.g., Breadth-First-Search (BFS) and Depth-First-Search (DFS)) to test the contiguity of every service area. If it cannot find the best PE-path that satisfies SA contiguity, then it returns "no solution found". PE then re-allots the best boundary graph-nodes on the best PE-path (line 8). This process continues until the allotment is in line with the capacity of the service centers (line 2). Finally, the updated CCNVD with balanced service centers is returned (line 10).

Example of PE algorithm: Figures 3.5, 3.6, 3.7 and 3.8 show the execution of the PE algorithm. PE starts with NVD (Fig. 3.5a) and creates a PE-Graph (Fig. 3.5b). In this example, the service center with an excess is X and the service center with a deficit is Z. PE finds the best PE-path to traverse from X to Z (i.e., $X \rightarrow Y \rightarrow Z$) as well as the best boundary graph-nodes adjacent to other SAs. Figure 3.5c shows these nodes are C and E.

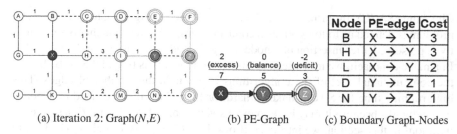

(a) Iteration 2: Graph(N,E)   (b) PE-Graph   (c) Boundary Graph-Nodes

Node	PE-edge	Cost
B	X → Y	3
H	X → Y	3
L	X → Y	2
D	Y → Z	1
N	Y → Z	1

Fig. 3.6 PE: Iteration 2

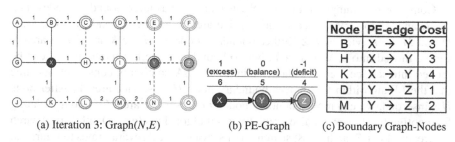

(a) Iteration 3: Graph(N,E)   (b) PE-Graph   (c) Boundary Graph-Nodes

Node	PE-edge	Cost
B	X → Y	3
H	X → Y	3
K	X → Y	4
D	Y → Z	1
M	Y → Z	2

Fig. 3.7 PE: Iteration 3

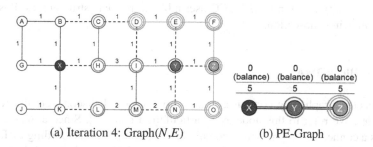

(a) Iteration 4: Graph(N,E)   (b) PE-Graph

Fig. 3.8 PE: Iteration 4 produces a CCNVD

Next, PE re-allots the best boundary graph-nodes (C and E) to the service centers in their adjacent SAs (Fig. 3.6a) and updates the PE-Graph (Fig. 3.6b). After three iterations, PE achieves a balanced allotment (Fig. 3.8b), having re-allotted nodes L and H from X to Y and nodes N and D from Y to Z, and the algorithm terminates. Figure 3.8a shows the resulting Capacity Constrained Network Voronoi Diagram.

3.2.2 PE-BTCC Algorithm

The main performance bottleneck of PE is the check for Service Area contiguity [21]. A contiguity checking algorithm is applied to test whether every re-allotment on the PE-path preserves SA contiguity. We refer to this testing as Service Area Contiguity

Checking (SACC). In formal terms, the Service Area Contiguity Checking (SACC) problem is defined as follows: given a connected graph SA, test whether SA contiguity is preserved after insertion of a node $n_{ins} \notin SA$ and removal of a node $n_{rem} \in SA$. In a naive approach, we may use graph traversal algorithms (e.g., DFS or BFS), but the computational cost of these algorithms is linear in the number of edges. This high running time for SACC makes it hard for PE to handle large sized transportation networks because it may have to extensively search SAs several times, in each iteration. In this section, we introduce Block Tree Contiguity Checking (BTCC) to reduce the computational cost of this step.

Consider the example in Fig. 3.7. Service area X has three boundary nodes (i.e., B, H, and K) to service area Y and service area Y has two boundary nodes (i.e., D and M) to service area Z. Either boundary node B, H, or K in service area X can be moved into service area Y, but no boundary node in service area Y can be moved into service area Z without violating the SA contiguity constraint. Assume that the PE algorithm first moves boundary node H to service area Y. After node H's insertion, we can move boundary node D into service area Z without violating SA contiguity. However, naive graph traversal algorithms (e.g., Depth-First-Search or Breadth-First-Search) must search all of service area Y several times to figure out which of its boundary nodes are movable to service area Z. To improve the efficiency of SA contiguity checking, PE-BTCC uses a block tree data structure, as illustrated in the following subsection.

3.2.2.1 Block Tree

A connected graph may contain a node whose removal disconnects the remaining nodes. We refer to this node as an articulation node [7]. Since a Service Area (SA) is a connected graph, it may contain articulation nodes. According to Tarjan's algorithm [19], we can create a DFS-spanning tree in linear time and detect these articulation nodes in constant time. A graph with no articulation nodes is called bi-connected or non-separable. A maximal bi-connected sub-graph of a graph is called a block [7]. Since our main focus is finding articulation nodes in the SA, we group these non-articulation nodes into blocks to simplify the representation of the DFS-spanning tree. We refer to this tree as a block tree [7, 19]. PE-BTCC uses a block tree that consists only of blocks and articulation nodes.

Figure 3.9 shows examples of DFS-spanning and block trees generated from Fig. 3.5a. The edges of the original graph can be divided into two types: tree edges and back edges. A tree edge belongs to the DFS-spanning tree itself; it connects a node to one of its descendants whereas a back edge connects a node to one of its ancestors. Figure 3.10a shows tree edges (solid lines) and back edges (dotted lines) based on the original graph.

After creating the DFS-spanning tree, we can easily see that every leaf node is a non-articulation node because its removal does not separate the SA. Consider the case of a non-leaf node that has children. If every child has a path to an ancestor of the non-leaf node with tree or back edges, then the non-leaf node is not an articulation

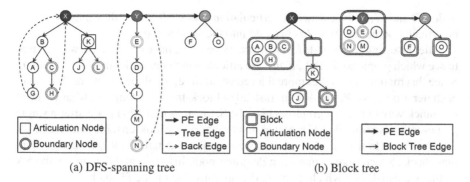

(a) DFS-spanning tree   (b) Block tree

Fig. 3.9 Example of DFS-spanning and block tree: Iteration 1

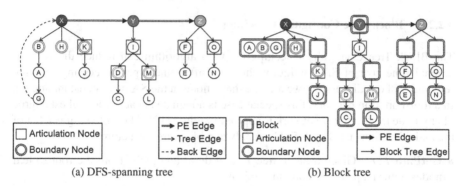

(a) DFS-spanning tree   (b) Block tree

Fig. 3.10 Example of DFS-spanning and block tree: Iteration 3

node because these paths create a bi-connected sub-graph that includes the non-leaf node and its children. However, if a child has no path to any ancestor of the non-leaf node, it becomes an articulation node because this child becomes orphaned from the root after removal of its parent node. In Fig. 3.9a, node K is a non-leaf node and has two children, J and L. Since neither child has a path to an ancestor of node K, node K removal separates the SA into three connected graphs (e.g., (A, B, C, G, H), (J), and (L)). However non-leaf node B has two children, both of whom have a path to an ancestor of node B (e.g., $A{\to}G{\to}X$ and $C{\to}H{\to}X$). Therefore, node B is not an articulation node and its removal does not separate the SA. We then group these non-articulation nodes into blocks and get a block-tree shown in Fig. 3.9b.

A block tree representation simplifies SA contiguity checking because it allows us to determine whether a node in the SA is an articulation or not in constant time. However, this approach has limited ability to handle node updates (e.g., node insertion and deletion). PE re-allots one graph-node to its adjacent SA along the best PE-path. During computation of the best PE-path, a new block tree needs to be created and maintained every time a graph-node is re-allotted to its adjacent SA. In the block tree in Fig. 3.10b (generated from Fig. 3.7a), service area Y has two boundary graph-nodes, D and M, both of which are articulation nodes. However, they may no longer be articulation nodes after H is inserted into the service area Y.

After inserting graph-node H into service area Y, we have to create a new block tree to see which graph-nodes in Y, if any, are articulation nodes under the new condition. Since this maintenance takes more than constant time, it will be a bottleneck point in each iteration of the PE algorithm. Instead of block-tree maintenance, what we need is a quick way to test if a graph-node in the SA is an articulation node after a single graph-node insertion. BTCC achieves this by exploiting look-up tables and the block tree structure. In the next subsection, we describe a detail about BTCC algorithm that checks SA contiguity under single graph-node insertion and deletion in the SA so that we can decide which graph-nodes are movable on the PE-path.

3.2.2.2   Block Tree Contiguity Checking

The Block Tree Contiguity Checking (BTCC) algorithm can reduce the computational time cost of SA contiguity checking after linear pre-processing. For the remainder of our discussion, we assume that a node in the SA has a constant degree. In a transportation network, this special case is adequate because the highest degree of any node in a highway network is approximately 4 [12]. The following is a list of the main functions used by BTCC to analyze a block tree structure.

- **GetPath**(T, n): Given a block tree T, a node n, and paths from the root to leaf nodes, return a path that contains node n.
- **GetLevel**(T, n): Given a block tree T and a node n, return a level of node n on the block tree.
- **Contains**(T, p, n): Given a block tree T, a path p, and a node n, return true if the path p contains the node n.
- **GetLCA**$(T, n_1, n_2, \ldots, n_k)$: Given a block tree T and k nodes $n_1, n_2, \ldots, n_k$, return the lowest common ancestor (LCA) of $n_1, n_2, \ldots, n_k$ that is located farthest from the root (i.e., the node at the highest level of T).
- **BTCC**(T, n_{ins}, n_{rem}): Given a block tree T, a node $n_{ins} \notin T$, and a node $n_{rem} \in T$, return true if node n_{rem} is an articulation node after node n_{ins} has been inserted into the block tree T.

The BTCC algorithm proceeds in three steps. First, it creates a block tree T according to the DFS algorithm [19]. Then it creates look-up tables for three functions: $GetPath()$, $GetLevel()$, and $Contains()$. For the $GetLCA()$ function, BTCC use the LCA algorithm [2, 4, 10, 15], which can answer in constant time after linear pre-processing of the block tree T. Finally, it calls $BTCC(T, n_{ins}, n_{rem})$ to see if node n_{rem} is an articulation node now that node n_{ins} has been inserted into T.

Consider again the block tree in Fig. 3.10b. Service area Y has two boundary nodes (e.g., D and M), both of which are articulation nodes. We first create a path table that contains all paths from the root to its leaf nodes (Fig. 3.11a). Next, we create look-up tables that return the value for three functions, $GetPath()$, $GetLevel()$, and $Contains()$ (Fig. 3.11b). We also need to create a look-up table for $GetLCA()$ according to the

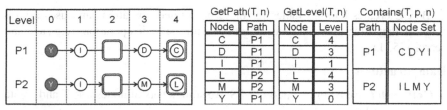

Level	0	1	2	3	4
P1	Y	→I	□	→D	→C
P2	Y	→I	□	→M	→L

GetPath(T, n)

Node	Path
C	P1
D	P1
I	P1
L	P2
M	P2
Y	P1

GetLevel(T, n)

Node	Level
C	4
D	3
I	1
L	4
M	3
Y	0

Contains(T, p, n)

Path	Node Set
P1	C D Y I
P2	I L M Y

(a) Path table for service area Y in block tree: Iteration 3

(b) Hash Tables

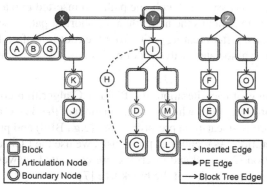

(c) Block graph after moving graph-node H to service center Y

Fig. 3.11 Example of path table and block graph: Iteration 3

LCA algorithm [2, 4, 10]. We assume that node H is moving from service area X to Y (Fig. 3.11c). When it does, node D is no longer an articulation node because insertion of H has created a bi-connected sub-graph (e.g., C, D, H, and I). We now show how to prove that node D is not an articulation node in constant time.

As we mentioned before, node D is an articulation node in the block tree for service area Y, and node H is a node inserted into service area Y (Fig. 3.11c). There are two incident nodes of H (i.e., C and I) and the lowest common ancestor (LCA) of these two nodes is I. Let P be a set of paths that connects the lowest common ancestor and incident nodes of n_{ins}. If P covers all children of n_{rem}, then node n_{rem} is not an articulation node because the insertion of node n_{ins} creates a bi-connected sub-graph that contains node n_{rem}. In this example, node D has only one child (C) and path $P1$ covers node C. Because the level of the lowest common ancestor (I) is 1 and the level of C is 4, the sub-path between I and C of $P1$ also covers node C. Therefore, node D is no longer an articulation node. This is tested simply by successively calling the predefined functions (e.g., *GetPath*(), *GetLevel*(), *Contains*(), and *GetLCA*()).

BTCC creates look-up tables for these predefined operations and uses a hash function that offers constant time performance for the basic operations (e.g., add and search) [5, 11].

Data Structures

- **GetLevel**(T, n): After computing BFS, we store levels of nodes $n \in T$ and create a look-up table.
- **GetPath**(T, n): After computing BFS, we create paths from the root to leaf nodes on the block tree T. The number of paths is bounded by the number of leaf nodes. Every node $n \in T$ is associated with one path and inserted into a look-up table. If more than one path contains the node, then the shortest path is selected.
- **Contains**(T, p, n): After creating paths from the root to leaf nodes on block tree T, every path p is associated with a set of nodes $n \in p$ and inserted into a look-up table.

At the beginning of each iteration, the PE-BTCC algorithm constructs look-up tables for these three functions which serve as an input of the BTCC algorithm. These look-up tables require linear-time pre-processing (e.g., BFS) and provide a constant time look-up operation. As mentioned previously, we use the LCA algorithm for the *GetLCA*() operation [2, 4, 10] and insert the pre-computed answers into a look-up table after linear pre-processing of the block tree [7].

Algorithm 2 Block Tree Contiguity Checking (BTCC) Algorithm (Pseudo-code)

Inputs:
- A block tree $T(N_{art}, N_{blk}, E)$ with a set of articulation nodes N_{art}, blocks N_{blk} and edges E.
- A node $n_{ins} \notin T$ that will be inserted into T.
- A node $n_{rem} \in T$ that will be removed from T.
- Look-up tables: *GetLevel*(), *GetPath*(), *Contains*(), *GetLCA*().
Outputs: Return true if n_{rem} is an articulation node after inserting n_{ins} into T
Steps:
1: **if** n_{ins} has only one incident in T and the incident is n_{rem}
2: **then** return true.
3: **if** n_{rem} is not an articulation node in T **then** return false.
4: $N_{incdt\_ins} \leftarrow$ (incident nodes of n_{ins}) $\in T$
5: $N_{ch\_rem} \leftarrow$ children of n_{rem}, $n_{lca\_incdt\_ins} \leftarrow LCA(N_{incdt\_ins})$
6: **if** $level(n_{lca\_incdt\_ins}) \leq level(n_{rem})$ **then** return true.
7: $P_{lca \rightarrow incdt} \leftarrow$ paths from $n_{lca\_incdt\_ins}$ to $n_{incdt\_ins} \in N_{incdt\_ins}$
8: **if** $P_{lca \rightarrow incdt}$ cover all $n_{ch\_rem} \in N_{ch\_rem}$ **then** return false.
9: return true.

Algorithm 2 presents the pseudo-code for the BTCC algorithm. First, BTCC checks whether node n_{ins} has only one incident on the block tree and whether the incident is the node n_{rem} (lines 1–2). If this is true, then node n_{rem} becomes an articulation node (Lemma 1). Next, it checks whether node n_{rem} is an articulation node on the block tree (line 3). If it is not, it is also not an articulation node after insertion of node n_{ins} (Lemma 2). BTCC then finds all incident nodes of n_{ins} and children of

n_{rem} as well as the lowest common ancestor of $n_{lca\_incdt\_ins}$ (lines 4–5). After that, it simply checks if the level of $n_{lca\_incdt\_ins}$ is greater than that of n_{rem}. If it is not, it returns true because no children have a path to an ancestor of n_{rem} (line 6). Next, it checks if these children (e.g., n_{ch_rem}) can be covered by the path from $n_{lca\_incdt\_ins}$ to all $n_{incdt\_ins} \in N_{incdt\_ins}$. If they can, it returns false (lines 7–8). Finally, it returns true after passing all the above criteria (line 9).

3.2.2.3 Proof of Correctness

The following lemmas prove the correctness of the Block Tree Contiguity Checking (BTCC) algorithm.

Lemma 1 *If a node n_{ins} inserted into a SA has only one incident in the SA and the incident is a node n_{incdt_ins}, then node n_{incdt_ins} becomes an articulation node.*

Proof After a node n_{ins} is inserted into the SA, the removal of n_{incdt_ins} disconnects the node n_{ins} from the SA. Therefore, node n_{incdt_ins} is an articulation node.

Lemma 2 *Except in the case of Lemma 1, given a SA, if a node n is not an articulation node in the SA, then it is also not an articulation after insertion of a node n_{ins} into the SA.*

Proof If a node n is not an articulation, the node is a part of a bi-connected sub-graph. Since the insertion of node n_{ins} does not decrease the connectivity of the SA, node n is not an articulation node.

Lemma 3 *Given a block tree T, an inserted node ($n_{ins} \notin T$), a set of its incident nodes ($n_1, n_2, \ldots, n_k \in T$), and a maximum node degree (n_{maxdeg}), GetLCA $(T, n_1, n_2, \ldots, n_k)$ gives an answer in $O(n_{maxdeg})$ time after linear-time pre-processing.*

Proof Given two nodes u and v, the Lowest Common Ancestor (LCA) algorithm gives an answer in constant time after linear-time pre-processing [5, 11]. The LCA function has commutative (e.g., $LCA(u, v) = LCA(v, u)$) and associative (e.g., $LCA(LCA(u, v), w) = LCA(u, LCA(v, w))$) properties. The number of incident nodes (e.g., k) is bounded by the maximum node degree (n_{maxdeg}). Therefore, $GetLCA(T, n_1, n_2, \ldots, n_k)$ can be computed with a time cost of $O(n_{maxdeg})$ by calling $LCA(\ldots LCA(LCA(n_1, n_2), n_3) \ldots, n_k)$.

Lemma 4 *After inserting a node n_{ins} and its incident edges E_{ins} into a block tree, an articulation node n is no longer an articulation node if its children have a path to one of the ancestors of node n.*

Proof Assume that after a node and its incident edges are added into the block tree, node n has children that have a path to one of its ancestors of node n with these new added edges E_{ins}. Then, even with removal of node n, its children maintain a connection with one ancestor of node n with edges E_{ins}. Therefore, node n is not an articulation node because its removal does not separate the graph.

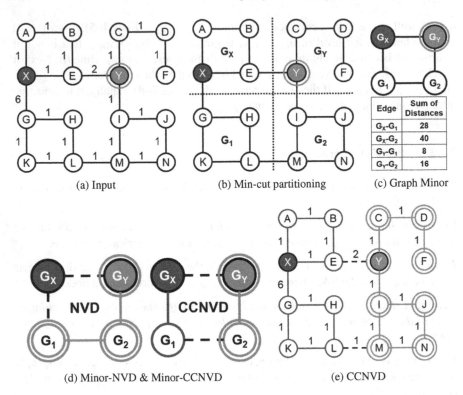

(a) Input   (b) Min-cut partitioning   (c) Graph Minor

(d) Minor-NVD & Minor-CCNVD   (e) CCNVD

Fig. 3.12 Example of PE algorithm using Graph Minor

3.2.3   PE-Minor Algorithm

Although BTCC reduces the computational cost for the bottleneck of the PE algorithm, it may be inapplicable for sizable road networks (e.g., USA road map). In this section, we introduce a Graph Minor approach that groups a set of graph-nodes and creates an initial partition to speed up the re-allotment. In graph theory, a minor of graph G can be formed from G by contracting edges and nodes [7]. PE-Minor uses a Graph Minor to group a set of graph-nodes and move multiple graph-nodes instead of one graph-node. As a pre-processing step, Graph Minor uses a balanced min-cut graph partitioning to decompose network G into connected components $c \in C$ and create a minor of G by contracting the edges and nodes in every $c \in C$. There are two reasons to use this method. First, a balanced min-cut partitions the network into same size sub-networks, making it possible to create minor nodes of the same load. Second, since a min-cut minimizes the number of edges between partitions, it can easily create a set of connected graph-nodes for every minor-node.

Example of PE-Minor algorithm: Figure 3.12 illustrates constructing of a CCNVD using Graph Minor. The input is a transportation network (14 graph-nodes

$(A, B, \ldots, N)$ and two service centers (X and Y)). Figure 3.12b illustrates balanced min-cut partitioning of the network. After partitioning, minor-nodes for the service centers are chosen (e.g., G_x and G_y). If more than one service center is located in the same minor-node, the set of graph-nodes in the minor-node is partitioned again with the NVD algorithm. Figure 3.12c shows a minor of the network after node and edge are contracted in every partition. As shown in the accompanying table, every minor-edge is associated with the sum of the shortest distances from a set of nodes in a partition to a service center. Next, the Minor-NVD and Minor-CCNVD are created with the PE algorithm. Finally, Fig. 3.12e shows the CCNVD after expanding of every minor node. The key idea behind the PE-Minor approach is to reduce the size of the network and move a set of graph-nodes through the PE-path.

3.3 Case Study with Brooklyn, NY Road Network

In our case study, we imagined a scenario in which victims of hurricane Sandy who needed to fuel up their cars were guided to the best gas station during the chaotic aftermath of the storm. For the transportation network, we used a Brooklyn, NY road map consisting of 7, 450 nodes and 22, 377 edges. We chose five gas stations and created a Network Voronoi Diagram (NVD), gas station allotment with the min-cost flow approach, and a CCNVD with the PE-algorithm. The number of allotted nodes for every gas station is represented by the size of the circles in the figures. For simplicity, we gave all gas stations equal capacity. In our analysis, the NVD (Fig. 3.13b) shows unbalanced allotments, which may lead to longer wait times for

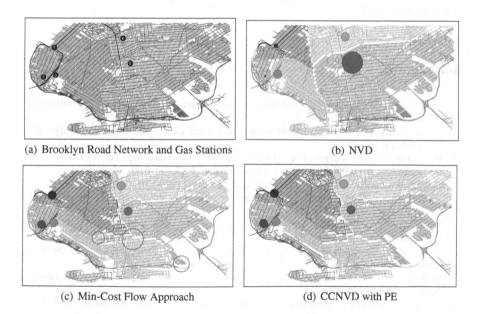

(a) Brooklyn Road Network and Gas Stations

(b) NVD

(c) Min-Cost Flow Approach

(d) CCNVD with PE

Fig. 3.13 Comparison of NVD, min-cost flow, and CCNVD with 5 gas stations

larger regions (e.g., gas station 4 (purple), gas station 5 (red), gas station 7 (rose)). Figure 3.13c shows that min-cost flow violates service area contiguity (the two green circles show areas of dis-contiguity). Figure 3.13d shows the CCNVD produced by our algorithm. We can see that the CCNVD is able to remove excesses and deficits so that all service centers are balanced in terms of the number of allotted nodes.

Given unbalanced allotments (e.g., NVD), the PE approach smoothly expands (or shrinks) service areas to meet capacity constraints. The PE-BTCC algorithm can achieve a significant computational performance gain over the PE algorithm because in each iteration, BTCC quickly tests the connectivity of the block tree with additional information. The PE-Minor can handle continental-sized transportation networks because it reduces the size of networks and speeds up the computation of the PE algorithm.

3.4 Summary

This chapter explored the problem of creating a Capacity Constrained Network Voronoi Diagram (CCNVD). An important potential application of CCNVD is promoting transportation resiliency after a disaster. Creating a CCNVD is challenging because of the large size of the transportation network and the constraint that service areas must be contiguous in the graph to simplify communication of service center allotments. The chapter presented Pressure Equalizer (PE) approaches for creating a CCNVD that meets the capacity constraints of service centers while maintaining the contiguity of service areas assigned to those centers. A case study using NY road map was presented.

References

1. Ahuja R et al (1993) Network flows: theory, algorithms, and applications. Prentice Hall, Englewood Cliffs
2. Alstrup S et al (2002) Nearest common ancestors: a survey and a new distributed algorithm. In: SPAA, pp 258–264
3. Barth D et al (2006) A degree bound on decomposable trees. Discrete Math 306(5):469–477
4. Bender MA et al (2000) The LCA problem revisited. In: LATIN'00: proceedings of the 4th Latin American symposium on theoretical informatics. Springer, London, pp 88–94
5. Brodnik A et al (1999) Membership in constant time and almost-minimum space. SIAM J Comput 28(5):1627–1640
6. Daskin M (2013) Network and discrete location: models, algorithms, and applications. Wiley-interscience series in discrete mathematics and optimization. Wiley, Hoboken
7. Diestel R (2005) Graph theory, vol 173, 4th edn. Graduate texts in mathematics. Springer, Heidelberg
8. Dyer M et al (1985) On the complexity of partitioning graphs into connected subgraphs. Discrete Appl Math 10(2):139–153
9. Erwig M (2000) The graph Voronoi diagram with applications. Networks 36(3):156–163

10. Fischer J et al (2006) Theoretical and practical improvements on the RMQ-problem, with applications to LCA and LCE. In: CPM'06: proceedings of the 17th annual conference on combinatorial pattern matching. Springer, Heidelberg, pp 36–48
11. Fotakis D et al (2005) Space efficient hash tables with worst case constant access time. Theory Comput Syst 38(2):229–248
12. Gastner MT et al (2006) The spatial structure of networks. Eur Phys J B 49(2):247–252
13. Győri E (1976) On division of graphs to connected subgraphs. In: Combinatorics, proceedings of fifth Hungarian Colloquium, Keszthely, 1976, vol 1, pp 485–494
14. Győri E (1981) Partition conditions and vertex-connectivity of graphs. Combinatorica 1(3):263–273
15. Harel D et al (1984) Fast algorithms for finding nearest common ancestors. SIAM J Comput 13(2):338–355
16. Johnson DS et al (1993) Network flows and matching: first DIMACS implementation challenge, vol 12. American Mathematical Society, Providence
17. Lovász L (1993) A homology theory for spanning trees of a graph. Acta Mathematica Academiae Scientiarum Hungarica 30(3–4):241–251
18. Okabe A et al (2008) Generalized network Voronoi diagrams: concepts, computational methods, and applications. Int J Geogr Inf Sci 22(9):965–994
19. Tarjan R (1971) Depth-first search and linear graph algorithms. In: 12th annual symposium on switching and automata theory, pp 114–121. doi:10.1109/SWAT.1971.10
20. Yang K, Shekhar AH, Oliver D, Shekhar S (2013) Capacity-constrained network-Voronoi diagram: a summary of results. In: International symposium on spatial and temporal databases. Springer, pp 56–73
21. Yang K, Shekhar AH, Oliver D, Shekhar S (2015) Capacity-constrained network-Voronoi diagram. IEEE Trans Knowl Data Eng 27(11):2919–2932

Chapter 4
Distance-Constrained k Spatial Sub-networks

4.1 Introduction

Given a graph and a set of spatial events (e.g., crime incidents, traffic collisions, etc.), The Distance-Constrained k Spatial Sub-Networks (DCSSN) problem finds k sub-networks that meet a distance constraint and maximize total number of spatial events covered by the sub-networks. Figure 4.1a shows an example input of DCSSN consisting of a graph with 14 nodes $(A, B, \ldots, N)$, 20 edges, and 25 locations of spatial events. Assume that $k = 3$ and the shortest path distance between nodes in a sub-network is at most 2. Figure 4.1b shows an example output of DCSSN. As can be seen, three sub-networks are returned, each indicated by a distinct line stye on its travel edges. The DCSSN problem is NP-hard (a proof is provided in Sect. 4.1.3). Intuitively, the problem is computationally challenging because of the large size of the transportation network and the distance constraint.

4.1.1 Application Domain

The DCSSN problem is important for critical applications such as identifying the most vulnerable areas within distance constraints. As an example, let us consider police patrol district design. Since crimes are unpredictable in the city, it is important to provide a reliable police service that responds quickly to calls. DCSSN identifies concentrations of spatial events and allocates limited resources to districts that are in the highest risk areas. DCSSN minimizes dispatch time to incidents by providing compact sub-networks. DCSSN can also be applied to discover groups of spatial locations that effectively interact and communicate with one another because it ensures the travel time between two locations is within the constraint. Possible examples of such situations are provided in Table 4.1.

© Springer International Publishing AG 2017
K. Yang and S. Shekhar, *Spatial Network Big Databases*,
DOI 10.1007/978-3-319-56657-3_4

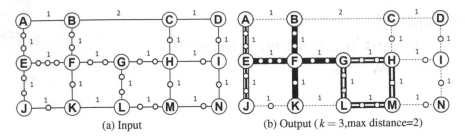

(a) Input   (b) Output ($k = 3$, max distance=2)

Fig. 4.1 Example of the Input and Output of DCSSN

Table 4.1 Possible applications of DCSSN

Application	Benefit of DCSSN
Police Patrol District	Manage police patrol to concentrate on high-crime areas
Fire Prevention	Identify highly vulnerable areas and watch for early signs of fire
Disease Surveillance and Response	Detect and monitor topological high-risk areas to prevent the spread of infectious diseases
Road Traffic Control	Identify high traffic accident areas for special attention and enforcement

4.1.2 Problem Definition

In the formulation of the DCSSN problem, a transportation network is represented and analyzed as a graph composed of nodes and edges. Each node represents a spatial location in geographic space (e.g., road intersections) and each edge between two nodes represents a road segment and has a travel distance. A spatial event has a spatial location on an edge. The $DCSSN(N,E,D,I,k,d_{max})$ problem is defined as follows:

Input: A transportation network G with

- a set of nodes N and a set of edges E,
- a set of non-negative integer lengths of edges $D : E \rightarrow \mathbb{Z}_0^+$
- a set of spatial event locations I,
- the number of sub-networks k, and
- the distance constraint d_{max}

Output: A set of Distance-Constrained k Spatial Sub-Networks SG_k

Objective:

- Maximize total number of spatial events covered by k sub-networks (SG_k).

Constraints:

- Distance Constraint: The shortest path distance between two nodes in $sg \in SG_k$ should be no greater than d_{max}.

4.1.3 Problem Hardness

The NP-hardness of DCSSN follows from a well know result about the NP-hardness of the maximum clique problem.

Theorem 3 *The DCSSN problem is NP-hard [28].*

Proof The NP-hardness of DCSSN can be proved by reduction from a well known NP-complete problem, the maximum clique problem (MCP). Given a graph G, MCP finds the largest clique. Let $A = (N, E)$ be an instance of MCP, where N is a set of nodes, and E is a set of edges. Let $B(N, E, D, I, k, d_{max})$ be an instance of the DCSSN problem, where N is a set of nodes, E is a set of edges, D is a set of distances of E, I is a set of spatial events in E, k is the number of sub-networks, and d_{max} is the distance constraint. Let k be 1 and let d_{max} be 1. Then it is easy to show that the instance of MCP is a special case of DCSSN, where every edge has a distance of 1 and contains exactly one spatial event $i \in I$. A is constructed from B in polynomial-bounded time. Therefore, the proof is complete.

4.1.4 Literature Review

Circle covering problems have been studied to find complete and partial spatial covering on a geometric space [1, 5, 6, 10, 11, 14, 22]. However, geometrical approaches (e.g., Euclidean distance) are not ideal for spatial networks [3, 23]. Metric k-center problems can find complete coverage of spatial events, which minimizes the longest edge between the center and spatial event locations [12, 15, 16]. However, these approaches are not designed to honor distance constraints between two nodes in a sub-network and cover all spatial events, leading to a limitation of the detection of distance-constrained spatial sub-networks. Clustering methods have been widely used in related research on a partial coverage problem [2, 9, 13, 26]. However, these methods do not consider distance constraints to build sub-networks. There exists a significant body of research on spatial network analysis. The K-function has been applied to spatial networks to analyze the distribution of events and detect clusters [19, 20, 25, 27]. Scan statistics has been used to detect anomalies on networks [17, 21]. The concept of spatial auto-correlation has been used to analyze the correlation between two variables on spatial networks [4, 7]. Network kernel density estimation analyzes the probability distributions of events and provides visual patterns of relative density on spatial networks [8, 18]. There is a network-based variable-distance clumping method that can discover multi-scale network-based clumps [24]. None of these approaches, however, consider distance constraints in its problem formulation. The Distance-Constrained k Spatial Sub-Networks (DCSSN) approach presented here honors distance constraints while also maximizing coverage of spatial events.

4.1.5   Outline of the Chapter

The rest of the chapter is organized as follows: Sect. 4.2 describes an algorithm that
can create a DCSSN. Section 4.3 presents a case study using Chicago crime datasets.
Finally, Sect. 4.4 summarizes the chapter.

4.2   Algorithm for Distance-Constrained k Spatial Sub-networks

This section presents the Rooted Sub-Graph with the Nearest Neighbor Distribution
(RSG-NND) algorithm to the DCSSN problem. The algorithm begins by constructing a set of rooted sub-graphs (RSG) and the nearest neighbor distribution (NND)
functions, and then it finds an attractor node located in the area with the highest
event density. The key idea in RSG-NND is to construct a data-structure to index the
nearest-neighbor events and assign the highest nearest-neighbor weighted edges to
the attractor node to maximize the coverage of spatial events. This process creates
sub-networks under distance constraints and iteratively finds dense sub-networks
in topological space.

The RSG-NND algorithm has three main steps: (1) construction of the rooted
sub-graphs and the nearest neighbor distribution functions, (2) assignment of spatial
events to sub-networks that maximizes the coverage of spatial events and honors the
distance constraint, and (3) update of the nearest neighbor distribution functions.

In the first step, RSG-NND creates the rooted sub-graph from every node ($r \in N$)
and constructs the nearest neighbor distribution (NND) function. The idea of RSG
is to create a sub-graph induced by all nodes within distance d of the root r.

Definition 2 The rooted sub-graph RSG(r,d) is the sub-graph of $G(N, E)$ spanned
by all nodes of G at a distance at most d from the root r.

The nearest neighbor distribution (NND) function can be represented by

$$NND(r, d) = \sum_{e \in RSG(r,d)} \#events(e), \qquad (4.1)$$

where r is the root of RSG, d is the distance constraint, and $\#events(e)$ is the number
of spatial events in edge e.

Figure 4.2 shows an example of the NND function. Figure 4.2a illustrates the input
with a transportation network (14 nodes, 20 edges, and 25 spatial events). Every edge
is associated with a distance (e.g., travel time), as indicated by the number displayed
alongside it. Consider NND for node F. Figure 4.2b shows the data-structure to index
the accumulated counts of nearest-neighbor events (i.e., $\#event(e)$) from node F. For
example, 17 events are located within the distance of 2 from node F. This model,
which we refer to as a distance-aggregated list, allows for storage of nearest neighbor

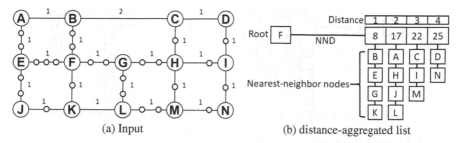

(a) Input　　　　　　　　　　　(b) distance-aggregated list

Fig. 4.2 Creation of nearest neighbor distribution function of node F

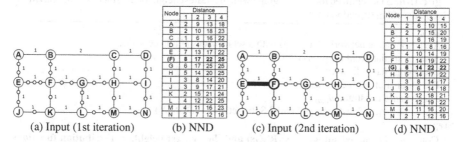

(a) Input (1st iteration)　　　(b) NND　　　(c) Input (2nd iteration)　　　(d) NND

Fig. 4.3 RSG-NND ($k = 3$, $d_{max} = 2$) (1st to 2nd iteration)

nodes in every distance. This is useful for NND queries as it updates the NND in linear time [28].

In the second step, RSG-NND examines the accumulated counts of events (i.e., values of NND) and chooses the node with the highest values. We refer to this node as an attractor node. Intuitively, the node with the highest values in NND has the largest expectation to become the attractor node which absorbs the next attractor node to form the DCSSN. After choosing the attractor node a, RSG-NND assigns the highest weighted edge (e.g., the number of spatial events) to the node a under the distance constraint (i.e., d_{max}). This process creates a sub-network that maximizes the coverage of spatial events under the constraint.

Figure 4.3 shows the process of selection of the attractor node and assignment of spatial events to the node. Figure 4.3b shows the NND for the network in Fig. 4.3a. Consider $d_{max} = 2$ and $k=3$. In this example, node F has the highest weight in the closed interval between 1 and 2 (i.e., [1, 2]). After the selection of the attractor node (i.e., F), RSG-NND scans all edges under the distance constraint and allots the highest weighted edge (i.e., EF) to node F (see Fig. 4.3c).

In the third step, RSG-NND removes all the spatial events on the allotted edge (i.e., EF) and updates the NND function. In order to minimize the computational cost for the update, RSG-NND uses two key ideas. First it constructs the distance-aggregated list and updates NNDs in linear time [28]. Second it updates only the part of nodes affected by the removal of spatial events (i.e., the updates are applied only to nodes which are reachable from the allotted edge within the distance constraint).

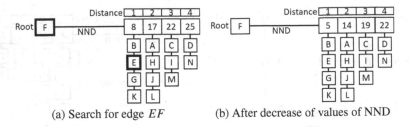

(a) Search for edge *EF*   (b) After decrease of values of NND

Fig. 4.4   Update of distance-aggregated list

Algorithm 3 Generalized rooted sub-graph with the nearest neighbor distribution function (RSG-NND) Algorithm (Pseudo-code)

Inputs:
- A transportation network ($G(N, E)$) with a set of nodes N and edges E.
- A set of spatial event locations I on E
- Every edge has a distance $d(e)$
- The number of sub-networks k
- The distance constraint d_{max}

Outputs: Distance-Constrained k Spatial Sub-Networks ($DCSSN$)

Steps:

1: Construct **the rooted sub-graphs (RSGs) and the nearest neighbor distribution functions (NNDs).**
2: $DCSSN \leftarrow \emptyset$
3: **while** $|I| > 0$ **do**
4:    Find an attractor node a in the range between $d_{max}/2$ and d_{max}.
5:    Assign the highest weighted edge $e_h \in E$ to a to maximize the coverage of spatial events under the distance constraint (i.e., d_{max}).
6:    $DCSSN \leftarrow DCSSN \cup e_h$
7:    Remove spatial events $i \in I$ from G.
8:    Update NNDs according to removal of spatial events i.
9: **end while**
10: return $DCSSN$ (i.e., k sub-networks that maximize coverage of spatial events.)

Figure 4.4 shows the update of NND for node F. After removing three spatial events on edge EF, RSG-NND decreases the value of 3 from d = 1 to 4. The result of the update is shown in Fig. 4.3d. This process continues until all the spatial events are removed and terminates in $O(m)$ iterations, where m is the number of edges.

Algorithm 3 presents the pseudo-code for a generalized version of RSG-NND. First, RSG-NND creates the rooted sub-graphs and the nearest neighbor distribution functions in the closed interval between $d_{max}/2$ and d_{max} (lines 1). Then, it chooses the highest NND node as an attractor node (Line 4). After that, it sorts edges in RSG based on the weights and assigns the highest weighted edge to the attractor node under the distance constraint (i.e., within distance d_{max}) (Line 5). If this sub-network can be added to an existing sub-network under the distance constraint, two individuals are combined (Line 6). If it cannot be combined with an existing sub-network, it becomes a new sub-network (Line 6). The algorithm then removes the spatial events

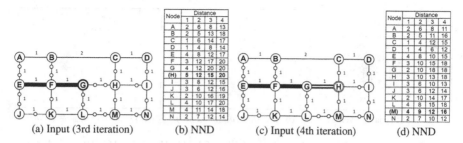

Node	Distance			
	1	2	3	4
A	2	6	8	13
B	2	5	13	18
C	1	6	14	17
D	1	4	8	14
E	4	8	12	17
F	3	12	17	20
G	4	12	20	20
(H)	5	12	15	20
I	3	8	12	15
J	3	6	12	16
K	2	10	16	19
L	4	10	17	20
M	4	11	14	18
N	2	7	12	14

Node	Distance			
	1	2	3	4
A	2	6	8	11
B	2	5	11	16
C	1	4	12	15
D	1	4	6	12
E	4	8	10	15
F	3	10	15	18
G	2	10	18	18
H	3	10	13	18
I	3	6	10	13
J	3	6	12	14
K	2	10	14	17
L	4	8	15	18
(M)	4	9	12	16
N	2	7	10	12

(a) Input (3rd iteration)   (b) NND   (c) Input (4th iteration)   (d) NND

Fig. 4.5 RSG-NND ($k = 3$, $d_{max} = 2$) (3rd to 4th iteration produces a DCSSN)

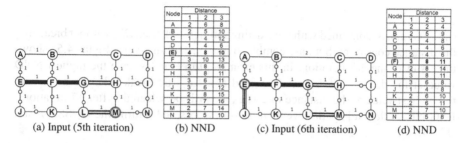

Node	Distance		
	1	2	3
A	2	6	8
B	2	5	10
C	1	4	12
D	1	4	6
(E)	4	8	10
F	3	10	13
G	2	8	16
H	3	8	11
I	3	6	11
J	3	6	12
K	2	8	15
L	2	7	16
M	2	7	14
N	2	5	10

Node	Distance		
	1	2	3
A	2	2	4
B	2	5	9
C	1	4	8
D	1	4	6
E	2	4	6
(F)	3	8	11
G	2	8	14
H	3	8	11
I	3	6	8
J	1	4	8
K	2	6	10
L	2	6	11
M	2	7	10
N	2	5	8

(a) Input (5th iteration)   (b) NND   (c) Input (6th iteration)   (d) NND

Fig. 4.6 RSG-NND ($k = 3$, $d_{max} = 2$) (5th to 6th iteration produces a DCSSN)

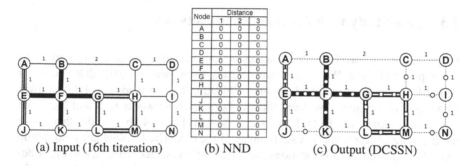

Node	Distance		
	1	2	3
A	0	0	0
B	0	0	0
C	0	0	0
D	0	0	0
E	0	0	0
F	0	0	0
G	0	0	0
H	0	0	0
I	0	0	0
J	0	0	0
K	0	0	0
L	0	0	0
M	0	0	0
N	0	0	0

(a) Input (16th titeration)   (b) NND   (c) Output (DCSSN)

Fig. 4.7 RSG-NND ($k = 3$, $d_{max} = 2$) (17th iteration produces a DCSSN)

in the network G (Line 7) and updates the NNDs (Line 8). This process continues until all the spatial events are removed. Finally, the DCSSN is returned (line 10).

Example of RSG-NND: Figures 4.3, 4.4, 4.5, 4.6 and 4.7 show the execution of the RSG-NND algorithm. First, the rooted sub-graphs and nearest neighbor functions are generated (Fig. 4.3b). In this example, node F becomes an attractor node (see Fig. 4.3b) and takes edge EF to construct the sub-network. Figure 4.3c shows the network after the allotment of edge EF to node F. Next, the NNDs are updated and the next attractor node (i.e., G) as well as edges with the highest weight (e.g., FG, GH, GL, and LM) are found. As a tie break rule, the edge connecting to the incident with the highest NND value (i.e., FG) will be allotted to node G. Since this allotment

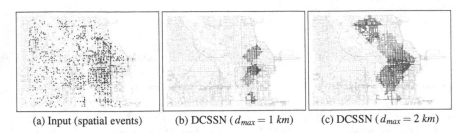

(a) Input (spatial events)   (b) DCSSN ($d_{max} = 1\ km$)   (c) DCSSN ($d_{max} = 2\ km$)

Fig. 4.8 Case Study: Chicago, IL road map ($|N| = 23,165$, $|E| = 35,161$, $|I| = 4,463$, and $k = 5$) (*Color* shows sub-networks) (Color figure online)

(i.e., *FG*) can be combined with the existing sub-network (i.e., *EF*) without breaching the distance constraint, both are combined into one sub-network. Figure 4.5c shows the result of the *4th* iteration. In this example, three nodes have the highest NND value at the distance of 1 (i.e., *E*, *L*, and *M*). As a tie break rule, the node with the highest value at the distance of 2 (i.e., *M*) will be selected. After 17 iterations, RSG-NND allots all spatial events to sub-networks and returns the DCSSN result (Fig. 4.7c).

4.3 Case Study with Chicago Road Network

In our case study, we imagined a scenario that finds five police patrol regions to cover potential risks. We used crime datasets from January 2015 that consist of 4,463 locations of crime incidents on a Chicago, IL road map (Fig. 4.8a) (https:// data.cityofchicago.org). We fixed the number of sub-networks to 5 and increased the distance constraint from 1 to 2 km (Fig. 4.8). Figure 4.8b, c show that crime incidents are concentrated in eastern areas of the city. The DCSSN detected within 1 km (Fig. 4.8b) covers 563 locations and the DCSSN detected within 2 km (Fig. 4.8c) covers 1,223 locations. As the distance (size) constraint increases, the sub-networks expand in an northerly and southerly direction.

4.4 Summary

We presented the problem of creating distance-constrained *k* spatial sub-networks (DCSSN). Creating a DCSSN is challenging because of the large size of the transportation network and the constraint that any two nodes in a sub-network must be within a predefined distance range. The chapter presented the rooted sub-graph with the nearest neighbor distribution function (RSG-NND) approach for creating a DCSSN to meet a distance constraint while maximizing the coverage of spatial events. A case study using Chicago crime datasets was presented.

References

1. Agarwal PK, Procopiuc CM (2002) Exact and approximation algorithms for clustering. Algorithmica 33(2):201–226
2. Aggarwal CC, et al (2013) Data clustering: algorithms and applications. CRC Press
3. Barthélemy M (2011) Spatial networks. Phy Rep 499(1):1–101
4. Black WR, Thomas I (1998) Accidents on belgium's motorways: a network autocorrelation analysis. J Transp Geogr 6(1):23–31
5. Carmi P, Katz MJ, Lev-Tov N (2007) Covering points by unit disks of fixed location. In: Algorithms and computation. Springer, pp. 644–655
6. Chazelle BM et al (1986) On a circle placement problem. Computing 36(1–2):1–16
7. Chun Y (2008) Modeling network autocorrelation within migration flows by eigenvector spatial filtering. J Geogr Syst 10(4):317–344
8. Downs J, Horner M (2007) Network-based kernel density estimation for home range analysis. In: Proceedings of the ninth international conference on geocomputation, Maynooth, Ireland
9. Ester M et al (1996) A density-based algorithm for discovering clusters in large spatial databases with noise. In Kdd, vol 96, pp. 226–231
10. Gandhi R, Khuller S, Srinivasan A (2004) Approximation algorithms for partial covering problems. J Algorithms 53(1):55–84
11. Ghasemalizadeh H, Razzazi M (2012) An improved approximation algorithm for the most points covering problem. Theor Comput Syst 50(3):545–558
12. Gonzalez TF (1985) Clustering to minimize the maximum intercluster distance. Theor Comput Sci 38:293–306
13. Han J, et al (2011) Data mining: concepts and techniques. Elsevier
14. Hifi M, M'hallah R (2009) A literature review on circle and sphere packing problems: models and methodologies. Adv Oper Res 2009
15. Hochbaum DS, Shmoys DB (1985) A best possible heuristic for the k-center problem. Math Oper Res 10(2):180–184
16. Hochbaum DS, Shmoys DB (1986) A unified approach to approximation algorithms for bottleneck problems. J ACM (JACM) 33(3):533–550
17. Marchette D (2012) Scan statistics on graphs. Wiley Interdisc Rev Comput Stat 4(5):466–473
18. Okabe A, Sugihara K (2012) Spatial analysis along networks: statistical and computational methods. Wiley
19. Okabe A, Yamada I (2001) The k-function method on a network and its computational implementation. Geogr Anal 33(3):271–290
20. O'Sullivan D, et al (2014) Geographic information analysis. Wiley
21. Priebe CE et al (2005) Scan statistics on enron graphs. Comput Math Organ Theor 11(3):229–247
22. Samet H (2006) Foundations of multidimensional and metric data structures. Morgan Kaufmann
23. Shekhar S, Chawla S (2003) Spatial databases: a tour. Prentice Hall, Upper Saddle River
24. Shiode S, Shiode N (2009) Detection of multi-scale clusters in network space. Int J Geogr Inf Sci 23(1):75–92
25. Spooner PG et al (2004) Spatial analysis of roadside acacia populations on a road network using the network k-function. Landscape Ecol 19(5):491–499
26. Tan P, et al (2005) Introduction to data mining. Addison-Wesley
27. Yamada I, Thill JC (2007) Local indicators of network-constrained clusters in spatial point patterns. Geogr Anal 39(3):268–292
28. Yang K (2016) Distance-constrained k spatial sub-networks: a summary of results. In: International conference on geographic information science. Springer, pp. 68–84

Chapter 5
Evacuation Route Planning

5.1 Introduction

Given a transportation network, a population, and a set of destinations, the goal of the Evacuation Route Planning (ERP) problem is to produce routes that minimize the evacuation time for the population. Evacuation planning is essential for ensuring public safety in the wake of man-made or natural disasters (e.g., terrorist acts, hurricanes, and nuclear accidents). The primary goal is to minimize evacuation time (i.e., the time from the start of the evacuation to last evacuee reaches destination) under the following two constraints: (1) evacuation routes should preserve capacity constraints of the transportation network, (2) computation time should be reasonable in order to respond to the urgent situation. Minimizing evacuation time is important because it minimizes exposure to potential hazards. Minimizing computation time is critical to enhance disaster response. The problem is challenging because of the large size of network data, the large number of evacuees, and the need to account for capacity constraints in the road network. Promising methods that incorporate capacity constraints into route planning have been developed but new insights are needed to reduce the high computational costs incurred by these methods with large-scale SNBD.

5.1.1 Application Domain

Hurricane Rita and the recent Tohoku tsunami that hit Japan are reminders that evacuation planning is an essential component of civic emergency preparedness. Ensuring the safety of all residents of a structure, city, or region during a disaster requires evacuation planning tools to produce the safest and most efficient route schedules for large scale road networks and populations within limited time constraints. Consider a hurricane evacuation planning problem. Low lying areas are especially at risk during major storms due to the likelihood of wide-spreading flooding. The speed and direction of a hurricane can change rapidly, so the threat to particular areas of

© Springer International Publishing AG 2017
K. Yang and S. Shekhar, *Spatial Network Big Databases*,
DOI 10.1007/978-3-319-56657-3_5

the coast may come up suddenly. Massive emergency evacuation from these areas brings more challenges for civic authorities due to the large and unpredictable shape of evacuation zones (EZs) along coastal areas. In 2005, the approach of hurricane Rita provoked one of the largest evacuations in U.S. history, resulting in three million evacuees. During the evacuation, the enormous number of people fleeing from the Houston area coupled with a number of shortcomings in exit routes for residents caused massive traffic jams. In 1992, Hurricane Andrew, the third most powerful storm to hit the Florida coast caused massive delays and major congestion [21].

5.1.2   Problem Definition

In the formulation of the Evacuation Route Planning (ERP) problem, a spatial network is represented and analyzed as a directed graph composed of nodes and edges. Each node represents a spatial location in geographic space (e.g., room, corridor, staircase, road intersection, and exit). Each edge between nodes represents a pathway (or road segment) and has a travel distance and capacity constraint. The problem of ERP is defined as follows:

Input: A transportation network with

- non-negative integer capacity constraints on nodes N and edges E,
- the total number of evacuees and their initial location, and
- location of evacuation destination

Output: An evacuation plan consisting of a set of origin-destination routes and a scheduling of evacuees on each route.

Objective:

- Minimize the computational cost of producing the evacuation plan
- Minimize the evacuation time.

Constraints:

- Edge travel time preserves FIFO (First-In First-Out) property.
- The scheduling of evacuees on each route observes the capacity constraints.
- Limited amount of computer memory

5.1.3   Literature Review

Over the last two decades there has been a considerable amount of research on route planning for evacuation zones and other event scenarios. Recent work on evacuation route planning can be divided into three categories: (1) Linear Programming (LP) methods that use a network flow problem to minimize the total evacuation time

[1, 9, 10, 17], (2) Simulation methods that model the evacuation route as individual movements [2, 9] or a traffic assignment problem [16], and (3) Heuristic methods that use an approximate optimization technique to minimize the computation time. The LP approach uses iterative algorithms (e.g., simplex or ellipsoid method) to minimize the cost function based on given constraints [1, 10, 17]. This approach requires using a static network model for a dynamic environment to generate optimal evacuation plans. Consequently, the transportation network needs to be transformed into a time-expanded graph (TEG) by constructing $T + 1$ copies of nodes and edges [5]. Unfortunately, the number of variables and iterations in this linear program is in general exponential in the size of the underlying network, limiting its usefulness for large scale networks [4]. Simulation methods use individual traveler behaviors or traffic assignment in greater detail including the interaction between vehicles. One problem with this model is that regulating individual movements or assigning traffic flow associated with Wardrop's equilibrium model [22] in emergency evacuation is very complicated, making it inappropriate for large evacuation scenarios. In contrast to previous approaches, heuristic approaches use approximate methods to find near-optimal solutions with minimizing the computational cost. A well known approach that falls in this category is the Capacity Constrained Route Planner (CCRP) [11, 14, 15]. These methods use time aggregated graphs and evaluate a shortest route with capacity constraints to find the evacuation route at each time step [7, 8]. It is useful for medium sized networks (e.g., 1-mile evacuation zone), but has a limitation for large scale networks (e.g., 50-mile evacuation zone) [23]. A common evacuation scenario exhibits dartboard network structure, leading it to be partitioned by dartboard network cuts (DBN-cuts) [23]. The Dartboard Network Cut based Approach to Evacuation Route Planning (DBNC-ERP) introduces the notion of dartboard network structure and explains how to organize and group evacuation routes [23]. It accelerates the routing algorithm by grouping multiple node-independent shortest routes to reduce the number of search iterations. For example, instead of a single-route shortest-path algorithm, the DBNC-ERP algorithm uses a node-independent shortest-paths algorithm to aggregate evacuees on different spatial locations and evaluate multiple evacuation routes at the same time.

5.1.4 Outline of the Chapter

This rest of the chapter is organized as follows: Sect. 5.2 describes algorithms for the ERP problem. Section 5.3 presents the experimental analysis of these algorithms. Finally, Sect. 5.4 summarizes the chapter.

5.2 Algorithms for Evacuation Route Planning

In a typical evacuation planning scenario, the spatial structure is represented and analyzed as a network model with non-negative integer capacity constraints on the nodes and edges. Given the initial locations and final destinations of a group of evacuees, the ERP problem aims to produces evacuation routes which can minimize the evacuation time.

Consider a simple Evacuation Route Planning (ERP) problem in a building, illustrated in Fig. 5.1a. Each room, corridor, staircase, and exit of the building is represented as a node and each pathway from one node to another node is represented as a directed edge. Every node has two attributes: maximum node capacity and initial node occupancy. For example, the maximum capacity of node $N1$ is 50, indicating that the node can hold at most 50 evacuees at any time step. The initial occupancy is shown to be 10, which means 10 evacuees prepare to move out of the node. Every edge has two attributes: maximum edge capacity and travel time. For example, the maximum edge capacity of $N4 \rightarrow N6$ is 5, indicating that at most 5 evacuees can traverse the edge. The travel time of this edge is 4, which means it takes 4 time steps to traverse the edge. Suppose we initially have 10 evacuees at node $N1$, 5 at node $N2$, and 15 at node $N8$, the ERP produces evacuation routes as shown in Fig. 5.1b. The objective of the ERP problem is to minimize the computational cost of producing the evacuation routes while minimizing evacuation time.

5.2.1 Capacity Constrained Route Planner Algorithm

The Capacity Constrained Route Planner (CCRP) algorithm solves the ERP problem using an iterative approach to search evacuation routes. In each iteration, the algorithm finds a route r with the earliest arrival time to any destination node from any source node, taking previous reservations and possible wait times into consideration. Then, CCRP computes the possible number of evacuees that can travel through r under capacity constraints. The possible number of evacuees can be defined by the minimum residual capacity of any edge in the route r. After that, CCRP updates (or reduces) the node and edge capacities on the route r for those evacuees. The algorithm terminates when all the evacuees finds routes to any of the destinations.

A key step in CCRP is to determine route r with the earliest arrival time. A naive approach to obtain the route r could involve executing Dijkstra's algorithm (generalized to work with edge capacities and travel times) for every of source and destination node followed by selection of the shortest path [14]. However, this becomes a major performance bottleneck and adversely affects the scalability of the algorithm [15]. This bottleneck is handled in CCRP by adding a super source node ss and edges with zero travel time and infinite capacity between ss and all other source nodes [15]. The pseudo-code for CCRP is shown in Algorithm 4.

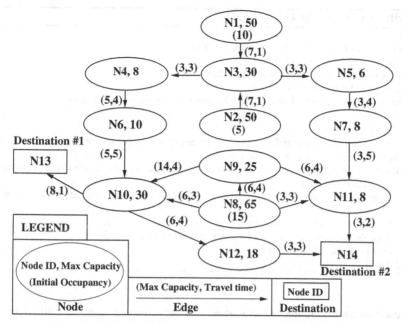

(a) Network Model for Simple Building Evacuation Scenario

Group of Evacuees			Route with Schedule	Start Time	Dest. Time
ID	Source	Group Size			
A	N8	6	N8(T0)–N10(T3)–N13	0	4
B	N8	6	N8(T1)–N10(T4)–N13	1	5
C	N8	3	N8(T0)–N11(T3)–N14	0	5
D	N1	3	N1(T0)–N3(T1)–N4(T4)–N6(T8)–N10(T13)–N13	0	14
E	N1	3	N1(T0)–N3(T2)–N4(T5)–N6(T9)–N10(T14)–N13	0	15
F	N1	1	N1(T0)–N3(T1)–N5(T4)–N7(T8)–N11(T13)–N14	0	15
G	N2	2	N2(T0)–N3(T1)–N5(T4)–N7(T8)–N11(T13)–N14	0	15
H	N2	3	N2(T0)–N3(T3)–N4(T6)–N6(T10)–N10(T15)–N13	0	16
I	N1	3	N1(T1)–N3(T2)–N5(T5)–N7(N9)–N11(T14)–N14	1	16

(b) Example of Evacuation Route Scheduling

Fig. 5.1 Building evacuation problem

The algorithm starts by adding edges between the super source node and every source node (Lines 1–2). These edges have a travel-time of 0 and infinity capacity. Then, the algorithm iterates (Lines 3–9) until there is no evacuee in any source node. Each iteration starts with finding the route r with earliest arrival time to any destination while considering the available capacities on nodes and edges. Then it determines the number of evacuees that the route can accommodate. This is done by computing all the minimum available capacities on r's edges (line 6). Finally, route

---

Algorithm 4 CCRP Algorithm (Pseudo-code)

Inputs:
  - A set of nodes N and edges E with capacity constraints C
  - Each edge $e \in E$ has a travel times t.
  - A set of source nodes S including initial evacuee occupancy O and
  - A set of destination nodes D
Outputs: Evacuation plan including route schedules of evacuees on each route r
Steps:
1: Add a super source node ss.
2: Add edges of 0 travel time and ∞ capacity between ss and all the source nodes.
3: **while** any source node $s \in S$ has evacuees **do**
4:    Find route r which has earliest destination arrival time using generalized version of Dijkstra's shortest path algorithm.
5:    Compute the minimum route capacity C_{min} with the edge and node capacity c along the route r.
6:    evacuee flow $f = min$(number of remaining evacuee at s in r , C_{min}).
7:    Reduce the node and edge capacity c along the route r using f.
8:    Remove evacuees from O using f.
9: **end while**

---

r is reserved for the number of evacuees, and capacities of its nodes and edges are updated (Line 7). The algorithm terminates when all the evacuees have been assigned a route.

5.2.2 Dartboard Network Cuts for Evacuation Route Planning Algorithm

CCRP finds a near-optimal evacuation plan with reduced computational cost [3, 11, 14, 15, 18]. This is useful for medium-sized transportation networks (e.g., 1-mile evacuation zone), but new insights are needed to handle large-scale network datasets (e.g., 3 million residents in a 30-mile evacuation zone) [23]. It is well known that a transportation network structure has an extremely important effect on the way vehicles move and exit evacuation zones (EZs) for emergency situations. For computational efficiency, it is important to exploit the underlying transportation network structure. In general, EZs are clearly defined in the evacuation plan documentation and often have a circular shape for nuclear evacuation plans and an elongated or irregular shape for hurricane evacuation plans. The network structure in EZs may be broken down into sections according to spatial movement patterns and analyzed independently for route evaluation. For example, circular EZs show 'outer first, inner last' flow patterns to evacuate the hazardous area, resulting in division of the transportation network in EZs into several rings. We define a dartboard network as a partitioned network according to the flow of vehicles such that all the vehicles in a single group reach the exit zone at the same time. The Dartboard Network Cuts for Evacuation Route Planning (DBNC-ERP) approach uses the notion of a dartboard network structure to model EZs (e.g., entire cities or coastal plains) in common evacuation scenarios. Instead of a single shortest route, DBNC-ERP evaluates multiple node-independent shortest routes at the same time to reduce iterations. The following subsection illustrates the basic concepts and the DBNC-ERP algorithm.

5.2.2.1 Basic Concepts

Spatial networks are represented and analyzed as a graph composed of nodes N and edges E. Every node N represents a spatial location in geographic space with a number of evacuees and a node capacity. Every edge E represents a connection between two nodes and has a travel time with an edge capacity. A sequence of nodes $n_1, n_2, n_3 \ldots, n_n$ is called a path (or route) if there is an edge between two consecutive nodes. A tree is an undirected graph in which any two nodes are connected by exactly one simple path. A forest is a disjoint union of trees. A set of paths from a source node s to a destination node d is a node-independent path if none of the paths share any nodes aside from s and d. In Fig. 5.2a, for example, there are two node-independent paths traversing from $S1$ to $T1$ ($S1 \rightarrow F \rightarrow B \rightarrow C \rightarrow T1, S1 \rightarrow K \rightarrow L \rightarrow H \rightarrow T1$). Figure 5.2b shows four node-independent paths based on two pairs of source and destination nodes ($S1 \rightarrow I \rightarrow T2, S1 \rightarrow N \rightarrow T2, S2 \rightarrow C \rightarrow T1, S2 \rightarrow H \rightarrow T1$). These node-independent paths are not necessarily unique so that there may be more than one way of choosing a set of independent paths.

Theorem 5.1 *Given a pair of nodes u, v, the upper bound of the number of node-independent paths is min(the degree of a node u, the degree of a node v).*

Proof Let m be $degree(u)$ and n be $degree(v)$. First, assume that $m \geq n$ and there exist n independent paths. When we add one more independent path, no incoming edge of n exists to obtain the independent path. Second, assume that $m < n$ and

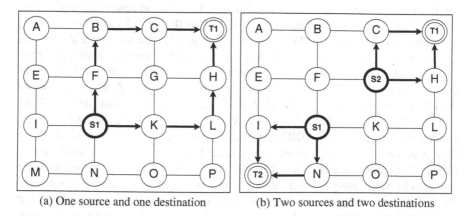

(a) One source and one destination   (b) Two sources and two destinations

Fig. 5.2 Node-independent routes in a grid-like network

there exists m independent paths. Again, we cannot add one more independent path because there is no available outgoing edge of m. Consequently, the maximum of node-independent paths is bounded by $min(degree(u), degree(v))$.

It is important to note that node-independent routes never share each other's capacity constraints at the same time during the route evaluation process.

5.2.2.2   Dartboard Network Structure

Spatial road networks were shaped in response to socioeconomic activities maximizing ease of navigation in the areas. The structures are neither trees nor perfect grids, but a combination of these structures that emerges from the social and constructive processes. The networks may be broken down into independent routes: most simply, routes that do not share any local parameters, such as node and edge capacity. These independent routes can partition evacuees and use discreet flows to compute the evacuation routes. Consider, for example, the grid-like road network in Fig. 5.2. Because independent routes do not share the capacity constraints of other roads in the network, one node-independent shortest path algorithm for such routes can minimize computation time. Unfortunately, many road networks have low degree intersections, making it hard to retrieve many node-independent routes. It is known that the mean degree of intersections in the US interstate road network is only about 2.86 [6]. By Theorem 5.1, we can retrieve at most 2.86 node-independent routes. One way to remedy the low degree issue is to use super nodes for source nodes and destination nodes. For instance, in Fig. 5.2b, $S1$ and $S2$ are grouped into a super source node S, and $T1$ and $T2$ are grouped into a super destination node T. Consequently, one node-independent shortest path algorithm for S and T can compute four node-independent shortest routes as an increasing node degree of S and T.

The spatial network organization of a place has an extremely important effect on the way people move through space and time. Evacuation route planning involving large numbers of evacuees has a well defined evacuation zone (EZ) (e.g., entire cites or coastal plains), making it possible to find the spatial movement patterns on transportation networks. For example, in Fig. 5.3a, the circular EZ encloses 12 nodes and all the travelers on these nodes need to move out of the circle. The network has one unit of capacity with one unit of travel time and two evacuees per node. Figure 5.3b shows the network model for the EZ. The nodes inside the EZ are $(A, B, C, D, E, F, G, H, I, J, K, L)$ and the nodes outside the EZ are $(X1, X2, X3, X4, X5, X6, X7, X8, X9, X10, X11, X12)$. The nodes in the EZ are divided into two groups by a DBN-cut. The outside nodes (A, B, C, F, G, J, K, L) have a higher number of shorter routes available compared to the inside nodes (D, E, H, I), reflecting an "outer first, inner last" flow pattern. That means, after evacuees from boundary areas move out of outer boundary areas, there is a secondary wave of evacuees from inner regions into boundary areas that will prepare to move out of the EZ. To characterize our dartboard network structure, dartboard network structure is defined as a network organization partitioned according to Dartboard Network

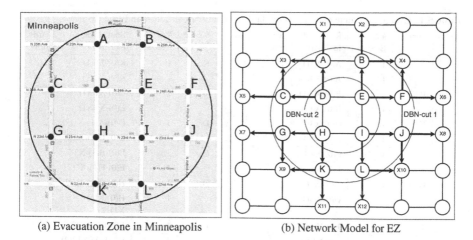

| (a) Evacuation Zone in Minneapolis | (b) Network Model for EZ |

Fig. 5.3 Dartboard network structure in road networks

Cuts (DBN-cuts), shown in Fig. 5.3b. To explain how this structure happens, look at the arrows in Fig. 5.3b. Given the EZ and number of evacuees, the evacuation plan needs to maximize the number of evacuees using the available shortest routes shown as arrows at each time step $t \in T$. As can be seen in Fig. 5.3b, the outer group (A, B, C, F, G, J, K, L) is poised to flee first from the EZ to maximize the number of evacuees, followed by the inner group (D, E, H, I), which takes its place and is similarly set to flee to maximize the number of evacuees again. We define a dartboard network cut (DBN-cut) as a cut in the flow of travelers in an evacuation network such that all the travelers in a single group are removed at the same time.

Theorem 5.2 *In a dart board network structure with FIFO property and a limited capacity constraint, a maximal dynamic flow algorithm for evacuation planning moves the outside nodes first and goes to inside nodes incrementally.*

Proof Assume that evacuees from inside nodes arrive at the destination before evacuees at outside nodes. This means that some outside nodes must have had to wait to exit the boundary area in order to make sure there is available capacity for evacuation of inside nodes. Otherwise, there would not have been enough capacity available for the inside nodes to exit the EZ through the outer boundary area. This implies an increase in the total time required to evacuate all people, thereby violating our objective to minimize the evacuation time.

Table 5.1 shows the results of an evacuation route plan with dart-board network structure. Each row in the table describes the schedule of a group of evacuees moving together to arrive at destinations at time step $t \in T$. During each iteration, the algorithm tries to group the node-independent routes and maximize the number of evacuees. For example, group 1 aggregates 16 node-independent shortest routes for 32 evacuees and group 2 aggregates 8 node-independent shortest routes for 16

Table 5.1 Evacuation route plan based on dartboard network structure in Fig. 5.3b

Group	Source	# of Evacuees	Route node Id(Time)	Arrival time	Group	Source	# of Evacuee	Route node Id(Time)	Arrival time
1	A	1	A(0)-X1(1)	1	1	K	1	K(0)-X9(1)	1
		1	A(0)-X3(1)				1	K(0)-X10(1)	
	B	1	B(0)-X2(1)			L	1	L(0)-X10(1)	
		1	B(0)-X4(1)				1	L(0)-X12(1)	
	C	1	C(0)-X3(1)		2	D	1	D(0)-A(1)-X1(2)	2
		1	C(0)-X5(1)				1	D(0)-C(1)-X5(2)	
	F	1	F(0)-X4(1)			E	1	E(0)-B(1)-X2(2)	
		1	F(0)-X6(1)				1	E(0)-F(1)-X6(2)	
	G	1	G(0)-X7(1)			H	1	H(0)-G(1)-X7(2)	
		1	G(0)-X9(1)				1	H(0)-K(1)-X11(2)	
	J	1	J(0)-X8(1)			I	1	I(0)-J(1)-X8(2)	
		1	J(0)-X10(1)				1	I(0)-L(1)-N12(2)	

evacuees. Note that each group aggregates evacuation routes on different spatial locations and reaches destinations at the same time.

Given the number of routes and the evacuation time, our principal objective is to maximize the number of evacuees for each time step $t \in T$. To achieve this end, an evacuation routing algorithm attempts to maximize the available shortest routes at each time step $t \in T$. Given this basic assumption, in each time step t, many node-independent routes may exist to maximize the number of available shortest routes. Next, we introduce our new algorithms to efficiently create DBN-cuts on evacuation networks.

5.2.2.3 DBNC-ERP

In this subsection, we describe the node-independent shortest paths approach for a dartboard network structure. A naive approach is to enumerate all available routes and remove node-dependent routes. However, the search space becomes exponential for combinations of the available multiple routes. General approaches for node-independent shortest paths construct a shortest-path tree and check the node dependency for each route [12, 19, 20]. The shortest-path tree of order n nodes has size $n - 1$, resulting in reduced search space by examining the boundary nodes [13]. In the ERP problem, there are many source nodes to traverse in order to reach destination nodes. Instead of a shortest-path tree, we can consider a shortest-path *forest* where each tree has a different root to handle multiple source nodes and iteratively choose node-independent shortest routes having available capacity. A forest of order n with k roots (or source nodes) has size $n - k$ since not every forest shares nodes. We allow shortest-path trees in the forest to share the same destinations because evacuees from different sources may reach the same destination. For each route, possible waiting time at each node is considered due to limited capacity constraint.

Algorithm 5 DBNC-ERP Algorithm (Pseudo-code)

Inputs:
- A set of nodes N and edges E with capacity constraints C
- Each edge $e \in E$ has a travel times t.
- A set of source nodes S including initial evacuee occupancy O and
- A set of destination nodes D

Outputs: Evacuation plan including route schedules of evacuees on each route r

Steps:

1: **while** any source node $s \in S$ has evacuees **do**
2:     Group all source nodes with a super source node ss and group all sink nodes with a super sink node sd.
3:     Construct a shortest forest from S to D based on ss and sd. Every tree can share the destination D.
4:     Find all routes R that are shortest node-independent paths from S to D.
5:     **for** $r \in R$ **do**
6:         Compute the minimum route capacity C_{min} with the edge and node capacity c along the route r.
7:         evacuee flow $f = min$(number of remaining evacuee at s in r , C_{min}).
8:         Reduce the node and edge capacity c along the route r using f.
9:         Remove evacuees from O using f.
10:    **end for**
11: **end while**

Algorithm 5 shows a way to identify the evacuation routes with node-independent shortest paths. The input for the pseudo code is an evacuation transportation network consisting of nodes, edges, source nodes, sink nodes, and capacity constraints. The output is an evacuation route schedule containing a sequence of nodes and edges. All source nodes and destination nodes are grouped by two super nodes to increase the node degree (Line 2). A shortest forest is constructed to find node-independent routes (Lines 3–4). After retrieving the node-independent routes, the capacity constraints are applied to these routes and available routes for evacuees are chosen (Lines 5–6). The next step is to reduce node and edge capacities along the routes (Lines 7–8) and repeat the above process until we finish finding the routes for all remaining evacuees.

5.3 Experimental Analysis

We evaluated the CCRP and DBNC-ERP algorithms for performance and scalability on a real world datasets.

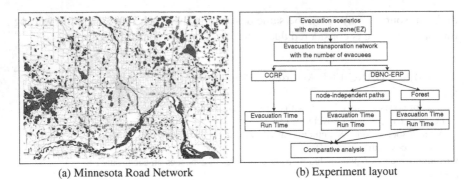

(a) Minnesota Road Network   (b) Experiment layout

Fig. 5.4 Experiment setup for ERP

5.3.1 Experiment Design

Figure 5.4b shows our experimental setup. Our dataset was a Minneapolis, MN road map consisting of 8, 868 nodes and 24, 126 edges, taken from TIGER/Line (http://www.census.gov). We tested two evacuation zones (EZs): one for a circular area and the other for a riverside area. Given the location of an incident and its scope R, we defined a circular EZ as the circular area centered at the incident with radius R and a riverside EZ (or coastal EZ) as the buffer area with R adjacent to rivers.

We tested three different approaches. The first two are CCRP [15] and DBNC-ERP with node-independent shortest paths (DBNC-ERP with NI). The third approach was DBNC-ERP with a shortest forest, as a candidate for DBNC-ERP to attempt to maximize the number of evacuees at each time step $t \in T$. The property of node-independent routes is easily exploited to reduce the computational time by reducing each iteration. However, the DBNC-ERP algorithm with node-independent shortest paths uses two route scans to evaluate the availability: one scan for node dependency and the other for capacity constraints. Intuitively, the forest for node-independent routes displays partially node-independent. We may relax the node-independent constraints and remove the node-dependency check for the forest. We call this method DBNC-ERP with a shortest forest. The strength of this approach is that it reduces route checking time when the number of available routes is large. The main disadvantage is that it may yield false-positive node-independent routes, which will be removed during capacity checking time.

The software was implemented in Java 1.7 with 1 GB memory run-time environment. All experiments were performed on an Intel Core i7-2670QM CPU machine running MS Windows 7 with 8 GB of RAM.

5.3.2 Experimental Results and Analysis

Three question were explored: (1) What is the effect of the number of evacuees? (2) What is the effect of the number of source nodes?, (3) What is the effect of the number of destination nodes?, (4) What is the effect of the size and shape of Evacuation Zone (EZ)?

Experiment 1: Effect of the number of evacuees:

The purpose of the first experiment was to evaluate the effect of number of evacuees on the performance of the algorithms. We fixed the number of source and destination nodes and multiplied the evacuees of each node. The experiment was done using networks of 246 source nodes, 109 destination nodes, and 1,847 nodes for a circular EZ. We incrementally increased the number of evacuees from 766,123 to 3,064,492. Figure 5.5a shows that the two DBNC-ERP approaches outperform CCRP. With increase of number of evacuees, the performance gap also increases. This is because the DBNC-ERP approaches group the node-independent routes to minimize the iterations. DBNC-ERP with a forest shows slightly better performance compared to DBNC-ERP with IN due to the longer node-dependency checking time. Figure 5.5b shows that all three algorithms were not distinguished in terms of evacuation time results. As the number of evacuees grows, the egress time increases.

Experiment 2: Effect of the number of source nodes:

The second experiment evaluated the effect of the number of source nodes on algorithm performance. We fixed the number of destination nodes and of evacuees. To increase the number of source nodes, source nodes were made to share evacuees with new source nodes. The experiment was done using networks of 109 destination nodes, 1,847 nodes for a circular EZ, and 766,123 evacuees. We incrementally increased the number of source nodes from 246 to 984. Figure 5.6a shows that number of source nodes has little effect on algorithm performance. Nevertheless, the two DBNC-ERP approaches run faster than CCRP.

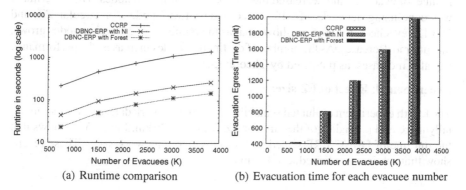

(a) Runtime comparison   (b) Evacuation time for each evacuee number

Fig. 5.5 Effect of the number of evacuees

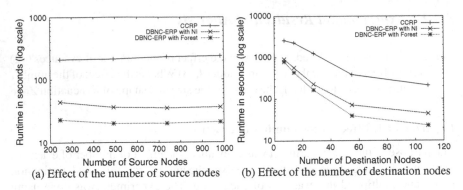

(a) Effect of the number of source nodes   (b) Effect of the number of destination nodes

Fig. 5.6   Effect of the number of source and destination nodes

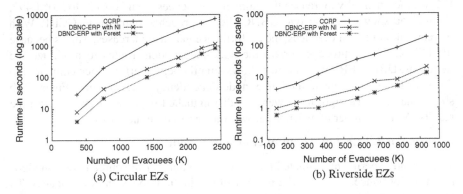

(a) Circular EZs   (b) Riverside EZs

Fig. 5.7   Effect of shape of EZ

Experiment 3: **Effect of the number of destination nodes**:

The third experiment evaluated the effect of the number of destination nodes on the performance of the algorithms. We fixed the number of source nodes, and the number of evacuees and decreased the number of destination nodes. The experiment was done using networks of 246 source nodes, 1,847 nodes for circular EZ, and 766,123 evacuees. Figure 5.6b shows that as the number of destination nodes grows, the runtime decreases. As the number of destination nodes increases, the performance gap also increases, as predicted by Theorem 5.1.

Experiment 4: **Effect of EZ size**:

The fourth experiment evaluated scalability for large network datasets. We incrementally increased the radius of the circular EZ from 5 to 30 km. Figure 5.7a shows that the runtime of DBNC-ERP scaled well to these large network sizes. These results show that runtime can be reduced by up to 80%

Experiment 5: Effect of EZ shape:

The fifth experiment evaluated the effect of the shape of the EZ on algorithm performance. If we put the destination nodes as boundary nodes of the EZ, a circular EZ will have a least possible number of boundary nodes due to its small surface area. If the EZ shows irregular shape (e.g., coastal areas), the number of boundary nodes increases as the surface area of the EZ increases. In our experiment, we chose evacuation zones along rivers in Minnesota and incrementally increased the length of the non-circular EZ to cover the entire length of the rivers. Results showed that the two DBNC-ERP approaches ran faster on the riverside EZ (Fig. 5.7b) than on the circular EZ (Fig. 5.7a). This is because the number of boundary nodes of the irregular shaped EZ is greater than of the circular EZ. Our results show that the runtime can be reduced by up to 90% when our approach is applied to irregularly shaped evacuation zones.

5.4 Summary

Evacuation route planning for large transportation networks is becoming increasingly important for dealing with man-made and natural disasters, such as hurricanes, terrorist acts, and nuclear accidents. An important component of evacuation planning methods is the ability to account for capacity constraints of the road network with manageable computational cost. In this chapter, we introduced a dartboard network structure to reflect an evacuee flow pattern for common evacuation scenarios by exploiting the spatial structure of the road network. Based on this structure, the DBNC-ERP algorithm partitions the network using dartboard network cuts (DBN-cuts) and groups source nodes in different spatial locations to maximize the number of evacuees. Experimental evaluation was present to show scalability of the algorithm on a real world datasets.

References

1. Ahuja R, Magnanti T, Orlin J, Weihe K (1993) Network flows: theory, algorithms and applications. Prentice Hall
2. Ben-Akiva M et al (2002) Development of a deployable real-time dynamic traffic assignment system: dynamit and dynamit-p users guide. Intelligent Transportation Systems Program, Massachusetts Institute of Technology
3. Evans MR, Yang K, Gunturi V, George B, Shekhar S (2015) Spatio-temporal networks: Modeling, storing, and querying temporally-detailed roadmaps. In: Space-Time Integration in Geography and GIScience. Springer, pp 77–108
4. Fleischer L, Skutella M (2007) Quickest flows over time. SIAM J Comput 36:1600–1630. Society for Industrial and Applied Mathematics, Philadelphia
5. Ford D, Fulkerson D (2010) Flows in networks. Princeton University Press, Princeton
6. Gastner M, Newman M (2006) The spatial structure of networks. Eur Phys J B-Condensed Matter Complex Syst 49:247–252. Springer

7. George B, Shekhar S (2008) Time-aggregated graphs for modeling spatio-temporal networks. In: Journal on data semantics XI, Springer, pp 191–212
8. George B, Kim S, Shekhar S (2007) Spatio-temporal network databases and routing algorithms: a summary of results. In: International symposium on spatial and temporal databases, Springer, pp 460–477
9. Hamacher H, Tjandra S (2002) Mathematical modelling of evacuation problems: state of the art. In: Pedestrian and evacuation dynamics. Springer, pp 227–266
10. Hillier F, Lieberman G, Hillier M (1990) Introduction to operations research. McGraw-Hill
11. Kim S, George B, Shekhar S (2007) Evacuation route planning: scalable heuristics. In: Proceedings of the 15th annual ACM international symposium on advances in geographic information systems. ACM, p 20
12. Kleinberg JM (1996) Approximation algorithms for disjoint paths problems. Ph.D. Dissertation, Department of CS, Massachusetts Institute of Technology
13. Korf R, Zhang W, Thayer I, Hohwald H (2005) Frontier search. J ACM (JACM) 52:715–748. ACM
14. Lu Q, Huang Y, Shekhar S (2003) Evacuation planning: a capacity constrained routing approach. In: International conference on intelligence and security informatics. Springer, pp 111–125
15. Lu Q, George B, Shekhar S (2005) Capacity constrained routing algorithms for evacuation planning: a summary of results. In: Proceedings of 9th international symposium on spatial and temporal databases. LNCS, vol. 3633. Springer, pp 291–307
16. Mahmassani H, Sbayti H, Zhou X (2004) Dynasmart-p: intelligent transportation network planning tool: version 1.0 users guide. Maryland Transportation Initiative, University of Maryland, College Park
17. Schrijver A (2003) Combinatorial optimization. Springer
18. Shekhar S, Yang K, Gunturi VM, Manikonda L, Oliver D, Zhou X, George B, Kim S, Wolff JM, Lu Q (2012) Experiences with evacuation route planning algorithms. Int J Geogr Inf Sci 26(12):2253–2265
19. Sidhu D, Nair R, Abdallah S (1991) Finding disjoint paths in networks. ACM SIGCOMM Comput Commun Rev 21:43–51 ACM
20. Suurballe J (1974) Disjoint paths in a network. Networks 4:125–145. Wiley
21. The New York Times (1992) HURRICANE ANDREW: When a Monster Is on the Way, 'It's Time to Get Out of Town'; In Texas, a Line of Cars 50 Miles Long, 26 Aug 1992. http://goo.gl/hq0EH
22. Wardrop J (1952) Some theoretical aspects of road traffic research. In: Proceedings of the institution of civil engineers, vol. 2, no. 1. Thomas Telford Ltd
23. Yang K, Gunturi VM, Shekhar S (2012) A dartboard network cut based approach to evacuation route planning: a summary of results. In: International conference on geographic information science. Springer, pp 325–339

Chapter 6
Storage Schemes for Spatio-Temporal Network Datasets

6.1 Introduction

A Spatio-Temporal Network (STN) can be defined as a graph $G = (N, E, T)$, where N is a set of nodes, E is a set of edges connecting two nodes, and every node and edge is associated with temporal information T (e.g., left-turn restriction and travel cost). STN datasets are becoming indispensable in societal applications, such as surface and air transportation management systems. One of the most appealing properties of these datasets is their ability to capture network attributes that vary over time. Consequently, STN datasets are usually massive in size and are accessed based on spatio-temporal movement patterns, making I/O efficient storage and access methods a significant challenge.

6.1.1 Application Domains

As an example, let us consider surface and air transportation management systems. The Federal Highway Administration [15] continuously records traffic data on major roads and highways across the United States using sensors, such as loop detectors, across the United States. Depending on the type of sensor, traffic levels are recorded as often as every minute or less. The Mobility Monitoring Program (MMP), started in 2000 by the Texas Transportation Institute, evaluated the use of sensors for traffic information around the United States. By 2003, the program was receiving traffic sensor data from over 30 cities and 3,000 miles of highway, with sensor readings occurring roughly every 30 s. These data are recorded 24 h a day, 365 days a year, resulting in millions of time steps per year for each sensor. In 2004, MMP published a report citing the need for better processing and storage of historical traffic data in order to benefit traffic management [15].

Another arena with a spatial-temporal network is the airline industry. Airlines connect thousands of destinations across the world through various routes between

© Springer International Publishing AG 2017
K. Yang and S. Shekhar, *Spatial Network Big Databases*,
DOI 10.1007/978-3-319-56657-3_6

airports. Maintaining accurate records of route performance is essential to evaluating and ensuring timely airline service, along with analyzing the potential causes and effects of delays. In order to measure route characteristics, such as average delay, each flight along the route is recorded with parameters such as flight time, departure time, landing time, etc. This flight information creates a spatio-temporal Network that allows historical queries to describe how often a flight is 'on time', 'late', 'very late', or 'excessively late'. Other more complex queries, such as how a delay on a particular route affects connecting flights, can also be analyzed with this data.

6.1.2 Basic Concepts

6.1.2.1 Spatio-Temporal Network and Lagrangian Path

A spatio-temporal network (STN) can be represented as a spatial graph with temporal attributes, composed of nodes (N), edges (E), and discrete time steps (t). Every node N represents a spatial location in geographic space. Every edge E represents a connection between two nodes and has temporal attributes associated with it. A sequence of nodes $n_1, n_2 \ldots, n_n$ is called a path (or route) if there is an edge between every two consecutive nodes.

Consider the snapshot model of a spatio-temporal network in Fig. 6.1. Spatial elements of the graph are a finite set of nodes (i.e., A, B, C, and D) and a finite set of edges connecting nodes (i.e., AB, AC, BD, and CD). In the figure, time steps 1, 2, 3, and 4 illustrate the progression of time in the network and the corresponding effect on edge attribute values (e.g., travel time).

STN queries can be described with a Lagrangian frame of reference [2]. We define a Lagrangian Path as a spatio-temporal network path where each edge is associated with time-varying attributes. For example, given two nodes, n_1 and n_2, and an edge e connecting the two nodes, a time-varying attribute such as travel time $d(e, t)$ can be associated with an edge $e(n_1, n_2)$ and interpreted as follows: if t is the departure time from node n_1, then $d(e, t) + t$ is the arrival time at node n_2. This implies that travel

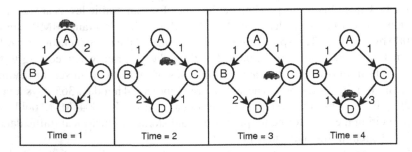

Fig. 6.1 Snapshot model of a spatio-temporal road network

Fig. 6.2 A spatio-temporal network as a time expanded graph

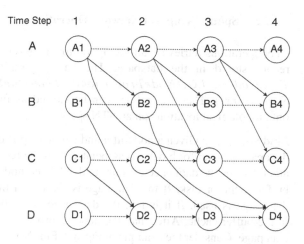

time on a Lagrangian path changes as time progresses. In Fig. 6.1, for example, at $t = 1$, edge AC has a travel time of 2. At $t = 2$, the edge value AC decreases to 1, indicating a reduced travel time for that section of the route. Consider traveling a route $A \rightarrow C \rightarrow D$. When starting at $t = 1$, it takes 3 time steps to reach D. However, when starting at $t = 2$, the same trip requires only 2 time steps. We refer to this query as route evaluation.

An STN dataset can be modeled as a time-expanded graph (TEG), that is, a spatio-temporal model which replicates each node along the time series such that a time-varying attribute is represented between replicated nodes [7, 13]. In a TEG, every node is associated with a temporal attribute, and every edge joining two nodes represents the spatio-temporal relation between the two nodes. Figure 6.2 illustrates a TEG for the spatio-temporal road network shown in Fig. 6.1. In Fig. 6.2, each edge contains the travel time needed to traverse the road segment. Given a start time of $t = 1$, a route $A \rightarrow C \rightarrow D$ consists of edge AC at $t = 1$ and edge CD at $t = 3$. The derived graph forms a directed, acyclic graph and neatly expresses a data access sequence for Lagrangian-connectivity.

6.1.2.2 Sub-node Data Structure of STN Datasets

Network representation is a crucial component of network analysis such as path computation and route evaluation. The studies in this book generally use an adjacent-list oriented representation due to its ability to support connectivity-based computation algorithms. For an STN, the node record format $Node(n, t)$ consists of a node-id n, a time step t, and time-varying information (e.g., turn restrictions and allowable waiting time) as well as outgoing edges for its successors (sub-nodes). Each edge has time-varying information such as travel time $d(e, t)$. We refer to this data structure as a sub-node data structure.

6.1.2.3   Spatio-Temporal Network Operations

STN operations retrieve spatio-temporal topology relationships between each node record stored in the database. The basic operations for STN datasets are: $FindNode()$, $LGetNodeTransition()$, $LGetOneSuccessor()$, and $LGetAll$ $Successors()$. These operations can be viewed as functions that map time steps onto topological locations on an STN space.

FindNode(n, t): Given a node-id n and a time step t, the operation searches for the data page pointer (or block-id) associated with the record from the secondary index and retrieves the data page containing the desired node record $Node(n, t)$ from the buffer cache or disk. If the data page is located in the buffer cache, the buffered data page is used. If it is not, the data page is fetched from the disk and stored in the buffer cache. After that, the node record $Node(n, t)$ is extracted within the data page. Consider the example in Fig. 6.2. FindNode($A, 1$) returns the node record $Node(A, 1)$.

LGetNodeTransition$(Node(n, t))$: Given a node record $Node(n, t)$, this operation retrieves the same node at the next time step, $Node(n, t + 1)$. The same-node transitions can be implemented by calling a $FindNode(n, t + 1)$. Alternatively, if the node record $Node(n, t)$ contains a data page pointer for the transition node, the operation goes directly to the data page and retrieves the node. This operation can be used to represent a wait on a spatial location. Consider the example in Fig. 6.2. LGetNodeTransition($Node(A, 1)$) returns the node record $Node(A, 2)$.

LGetOneSuccessor$(Node(n, t), n_s)$: Given a node record $Node(n, t)$ and a single successor-id n_s, this operation finds the successor's time step (t_s) from the node record $Node(n, t)$. Then it calls $FindNode(n_s, t_s)$ to retrieve $Node(n_s, t_s)$. Alternatively, if node record $Node(n, t)$ contains a data page pointer for successor n_s, the operation goes directly to the data page and retrieves $Node(n_s, t_s)$ without calling $FindNode(n_s, t_s)$. When the desired record is located in the buffer cache, no additional disk I/O is needed. Consider the example in Fig. 6.2. $LGetOneSuccessor(A1, B)$ returns one successor $B2$, while $LGetOneSuccessor$ $(A1, A3, B)$ returns $B2$, $B3$, and $B4$ for the interval $[A1, A3]$.

LGetAllSuccessors$(Node(n, t))$: Given a node record $Node(n, t)$, this operation finds all successor-ids (n_s) and their time steps (t_s) from the node record $Node(n, t)$. Then it calls $FindNode(n_s, t_s)$ for every successor. Alternatively, if the node record $Node(n, t)$ contains data page pointers for incident records, the operation goes directly to these data pages and retrieves all successors without calling multiple $FindNode(n_s, t_s)$. If the node records of successors can be made to share the same data page as $Node(n, t)$, we can significantly reduce the additional I/O cost of $LGetAllSuccessors()$. Consider the example in Fig. 6.2. $LGetAllSuccessors(A1)$ retrieves $B2$ and $C3$, whereas $LGetAllSuccessors$ $(A1, A3)$ examines the interval $[A1, A3]$ and returns $B2$, $B3$, $B4$, $C3$, and $C4$. The $LGetAllSuccessors()$ operation can be further optimized by exploiting the buffer cache. First, all data page pointers (or block-ids) for successors are searched from the secondary index or the node record, and node records of successors located in the buffer cache are retrieved first. After that, remaining record-pointers are sorted by data page pointer, and the relevant data pages are sequentially fetched from disk. This procedure can reduce the need for additional I/O even though there is only one buffer cache.

6.1.3 Problem Statement

The problem of Storing Spatio-Temporal Networks (SSTN) can be formalized as follows: given a spatio-temporal network (STN) and a set of STN operations, the objective is to find a storage scheme that minimizes the I/O costs of the STN operations. We formally define the problem as follows:

Input:

- A spatio-temporal network S
- A set of spatio-temporal operations O

Output:

- Data Partitioning of S, across data pages

Objective:

- Minimize data page access for operations in O

Constraints:

- S is too large for storage in main memory.
- Temporal-edge attribute information needs to be preserved.

In this chapter, we focus on two special cases of the SSTN problem: SSTN-G1S and SSTN-G$\forall$S. The objective of SSTN-G1S is to optimize the $LGetOne$ $Successor()$ operation for retrieving a single successor for a given node on an STN. The objective of SSTN-G$\forall$S is to optimize the $LGetAllSuccessors()$ operation for retrieving all successors for a given node on an STN. Both SSTN-G1S and SSTN-G$\forall$S problems are NP-hard and proofs are provided in Sects. 6.2.1.1 and 6.2.2.1. Intuitively, the two problems are computationally challenging because of the fixed data page size, the large size of STN datasets, and the constraint that data page access for the STN operation must be minimized.

6.1.4 Literature Review

Orthogonal partitioning methods, such as the longitudinal or snapshot method [3, 5, 16], are able to capture network connectivity orthogonally based on either

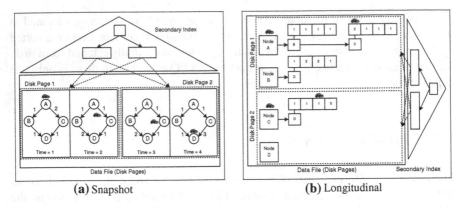

(a) Snapshot   (b) Longitudinal

Fig. 6.3 Storage representation and record format for orthogonal partitioning methods

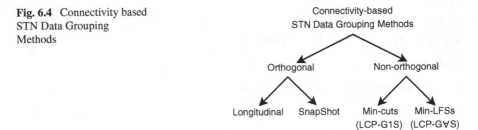

Fig. 6.4 Connectivity based
STN Data Grouping
Methods

space or time, as shown in Fig. 6.3. The snapshot method stores a topologically connected sub-graph for a given time instance into the same data page (Fig. 6.3a) whereas the longitudinal method stores temporally consecutive properties of a node (or edge) into the same data page (Fig. 6.3b). Current related work for storing and accessing STN data have relied on these orthogonal approaches [3, 5, 16]. However, the methods are not able to capture spatio-temporal connectivity that is both spatial and temporal in nature. To remedy this issue, non-orthogonal partitioning methods, such as SSTN-G1S and SSTN-G∀S, conceptualize the Lagrangian connectivity using a time-expanded graph and cluster the data records to reduce I/O costs [3, 5, 16]. A broader set of related work for the SSTN problem is summarized in Fig. 6.4. In the following section, we will review orthogonal approaches in more detail and illustrate efficient storage and access methods for SSTN-G1S and SSTN-G∀S.

6.1.5   Outline of the Chapter

The rest of this chapter is organized as follows: Sect. 6.2.1 describes the LCP-G1S approach using the concept of an LRatio. Section 6.2.2 describes the LCP-G∀S approach using the concept of a Lagrangian Family Set. Section 6.3 presents cost models for LCP-G1S and LCP-GAS. Section 6.4 presents an experimental analysis of the algorithms using real-world and synthetic datasets. Finally, Sect. 6.5 summarizes the chapter.

6.2 Lagrangian-Connectivity Partitioning Approaches for SSTN

This section describes Lagrangian-Connectivity Partitioning (LCP) approaches for the SSTN problem.

6.2.1 LCP-G1S for LGetOneSuccessor()

As the length of a time series in a spatio-temporal network increases, efficient execution of traversal queries requires a different approach for storing the data on disk. Traditional approaches partition networks based on nodes using some orthogonal emphasis (e.g., temporal or spatial). However, such methods do not work well with large STNs. Consider, for example, evaluating a route $(A \rightarrow C \rightarrow D)$ at a time step of 1 on the STN, shown in Fig. 6.2. Assume that one data page can store four nodes. With the orthogonal partitioning methods, such as snapshot (Fig. 6.5a) and longitudinal (Fig. 6.5b), whenever an edge is traversed $(A1 \rightarrow C3$ and $C3 \rightarrow D4)$, a disk I/O is needed to retrieve the data page containing the record for the next node. One naïve candidate to handle this issue is the aggregated time-stamped snapshot (ATSS) method that partitions the STN with static network connectivity and divides the time-series information into temporal chunks (Fig. 6.5c) [4]. The ATSS method can be seen as a trade-off between snapshot and longitudinal partitioning. ATSS was practical when the travel time was fairly uniform [4]. However, the main disadvantage is that it is not possible to determine the appropriate time interval parameter value to yield a better performance [4].

Figure 6.5d shows how LCP-G1S solves the SSTN-G1S problem. It is a non-orthogonal partitioning method that is optimized for retrieving a single successor for a given STN node. By performing a min-cut graph partitioning [10], we can create partitions based on single edge connectivity on an STN. As a result, spatio-temporally connected nodes can be collocated together on data pages. LCP-G1S results in more efficient I/O when calling a $LGetOneSuccessor()$ operation or queries composed of them [4]. In this example, with the LCP-G1S method, traversing from node $A1$ to $C3$ and then $C3$ to $D4$ requires only one data page as all relevant sub-node records are collocated on the same data page. In the next subsection, we define the SSTN-G1S objective function for minimizing the cost of $LGetOneSuccessor()$ and prove its NP-hardness.

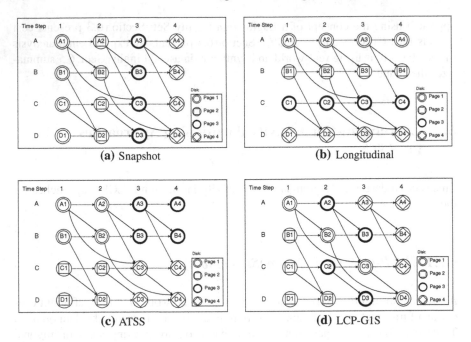

(a) Snapshot   **(b)** Longitudinal

(c) ATSS   **(d)** LCP-G1S

Fig. 6.5 STN partitioning methods

6.2.1.1 SSTN-G1S Objective Function and Problem Hardness

Traditional spatial networks use a connectivity ratio to measure predicated disk I/O [14]. We extended this connectivity ratio to formulate a spatio-temporal measurement, which we call a Lagrangian connectivity Ratio, or *LRatio*.

$$LRatio = \frac{Total\ number\ of\ unsplit\ Lagrangian\ edges}{Total\ number\ of\ Lagrangian\ edges} \quad (6.1)$$

The *LRatio* measures the connectivity along time and space in an STN. In Eq. (6.1), Lagrangian edges refer to edges connecting nodes through time, such as the edges displayed in a TEG. This metric ignores the 'wait' edges in a TEG, so maximizing the *LRatio* minimizes the disk I/Os for a $LGetOneSuccessor()$ operation.

Proposition 1 *The expected I/O cost of a $LGetOneSuccessor()$ operation is minimized by maximizing the LRatio.*

Proof Let $p(n)$ denote the id of the data page that stores node n. After calling $Find(n)$, the data page storing n is transferred into the buffer cache. Assume that $LGetOneSuccessor()$ retrieves one successor (e.g., s). Then the probability that the data page of the source node (i.e., n) is not the same as the data page of its successor (i.e., s) is $1 - Ratio$. Therefore, the expected number of pages can be represented by $1 - Ratio$. This implies the disk I/Os for $LGetOneSuccessor()$ is minimized by maximizing the $LRatio$.

The NP hardness of SSTN-G1S follows from a well-known result about the NP-hardness of the following balanced graph k-partitioning problem [1, 8]. Given a graph $G = (N, E)$, where N denotes a set of nodes and E a set of edges that can connect two nodes, the goal of the balanced graph k-partitioning problem is to partition N into equal sized parts $N_1, \ldots, N_k$ while minimizing edge-cuts. An edge is not cut if all nodes are in one partition, and cut exactly once otherwise [1].

Theorem 4 *The SSTN-G1S problem is NP-hard.*

Proof Given $STN(N, E, T)$, the SSTN-G1S problem partitions N into equal sized parts $N_1, \ldots, N_k$ while maximizing the $LRatio$. Consider the case $k = 2$. Then it is easy to show that the SSTN-G1S problem is equivalent to the balanced graph bi-partitioning problem, minimizing the edge-cuts between two partitions. Since the balanced graph bi-partitioning problem is constructed from SSTN-G1S ($k = 2$) in polynomial time, the proof is complete.

6.2.1.2 Algorithm for LCP-G1S

In this section, we introduce the LCP-G1S algorithm that produces a solution of SSTN-G1S. The LCP-G1S algorithm is based on a multi-way graph partitioning approach. First, STN is partitioned into equal-sized sub-networks by minimizing edge-cuts between sub-networks. Then, these sub-networks are stored as sub-node records. For efficiency, LCP-G1S uses a bulk load operation, which sorts these sub-node records along the block number and inserts them into data pages. Physically, sub-node data records are stored in data pages and a B+tree index is created to support retrieve operations.

Algorithm 7 shows one way to store node records using LCP-G1S. The input is a spatio-temporal network (STN) consisting of nodes, edges, and time values for each edge along with the physical page size for storage on disk. The output is a data file consisting of the data pages, containing records of node and edge information. Since the min-cut graph partitioning algorithm requires a predefined number of partitions, Line 1 estimates the number of pages (i.e., k) needed based on the size of the spatio-temporal network and the size of the data page. Line 2 constructs a time-expanded graph and Line 3 partitions this graph into equal sized sub-networks, thereby maximizing the $LRatio$. These sub-networks are then converted to sub-nodes in Line 6 and written to disk in Line 8.

6.2.2 LCP-G∀S for LGetAllSuccessors()

In this section, we introduce the LCP-G∀S algorithm that produces a solution of SSTN-G1S and explain key elements to design the algorithm in detail. We first

---

Algorithm 6 Pseudocode for the LCP-G1S algorithm

Inputs:

- A set of nodes N
- A set of edges E
- A set of travel times T
- Size of data page P

Outputs:

- Data file containing STN data across data pages (LCP-G1S-STN)

LCP-G1S
1: k = estimate num of data pages using V, E, T, and P
2: TEG = create time-expanded network from V, E, T
3: Part[] = run graph partitioning on edges in TEG using k
4: **for** each partition in Part [] **do**
5:    SN[] = create sub-nodes from partition
6:    **for** each sub-node in SN **do**
7:       RID = write sub-node to a data page
8:    **end for**
9: **end for**

---

point out that a solution of the SSTN-G1S problem is not entirely appropriate for the $LGetAllSuccessors()$ operation due to the structure of the set of successors. Consider the STN example in Fig. 6.6a. There are four parent nodes (e.g., $A1$, $B1$, $C1$, and $D1$) and four successor nodes (e.g., $A2$, $B2$, $C2$, and $D2$). We define a blocking factor as the number of node records per data page. In this example, we use a blocking factor of 4. We also assume that the data page for a parent node is located in the buffer cache before calling one of the STN operation: $LGetAllSuccessors()$ and $LGetOneSuccessor()$.

Figure 6.6b shows a solution of the SSTN-G1S problem and Fig. 6.6c shows a solution of the SSTN-G∀S problem. The dashed line represents the edge-cut that

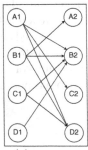

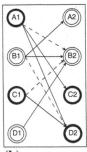

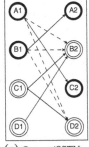

	I/O Cost			I/O Cost	
Node	LGetAllSuccessors()		Edge	LGetOneSuccessor()	
	SSTN-G1S	SSTN-G∀S		SSTN-G1S	SSTN-G∀S
A1	1	1	A1 B2	1	1
B1	1	1	A1 C2	0	0
C1	1	0	A1 D2	0	1
D1	0	0	B1 A2	0	0
Total	3	2	B1 B2	0	1
Avg.	0.75	0.5	B1 D2	1	1
			C1 B2	1	0
○	Page 1		C1 D2	0	0
○	Page 2		D1 B2	0	0
- - -►	Edge Cut		Total	3	4
			Avg.	0.333	0.444

(a) Input (STN)   **(b)** Output (SSTN-G1S)   **(c)** Output (SSTN-G∀S)   **(d)** I/O Costs of STN operations

Fig. 6.6 Comparison of SSTN-G∀S and SSTN-G1S

connects two nodes stored in different data pages. The additional disk I/Os for an STN operation is measured by the number of data page fetches for the operation call. Consider the $LGetAllSuccessors(A1)$ operation in Fig. 6.6b. $A1$ needs one more data page fetch to access $B2$, thereby causing one additional disk I/O for retrieving the set of successors.

The result of SSTN-G1S has three edge-cuts (Fig. 6.6b) whereas SSTN-G∀S has four edge-cuts (Fig. 6.6c). By contrast, Fig. 6.6d shows that SSTN-G1S requires 0.75 additional disk I/Os on average for the $LGetAllSuccessors()$ operation, whereas SSTN-G∀S requires 0.5 additional disk I/Os. Note that although the edge-cuts in SSTN-G∀S makes more edge-cuts, it still performs better than SSTN-G1S in terms of the I/O costs for the $LGetAllSuccessors()$ operation. This is true because SSTN-G1S ignores the grouping between a parent and its successor set. For this reason, the non-orthogonal partitioning approach requires a new objective function (i.e., LCP-G∀S) to optimize the $LGetAllSuccessors()$ operation. The basic idea of LCP-G∀S is to minimize the number of distinct data pages for a parent and its successors set. We refer to the set of data pages for a parent and its successors as a *Lagrangian Family Set (LFS)*.

6.2.2.1 SSTN-G∀S Objective Function and Problem Hardness

We define the SSTN-G∀S objective function for minimizing the cost of $LGetOne Successor()$. Then we prove its NP-hardness.

Proposition 2 *The expected I/O cost of a $LGetAllSuccessors()$ operation is minimized by minimizing $\sum_{n \in N} |LFS(n)|$, where $LFS(n)$ is a data page-id set for a node n and its successors.*

Proof Let $p(n)$ denote the id of the data page that stores node n. After calling $Find(n)$, the data page storing node n is transferred into the buffer cache. Clearly, the successors that share the same data page as node n do not require additional disk I/Os. On the other hand, the successors stored in $LFS(n) \setminus \{p(n)\}$ cause additional $|LFS(n)| - 1$ disk I/Os. This implies the disk I/Os for $LGetAllSuccessors()$ are minimized by minimizing $\sum_{n \in N} |LFS(n)|$.

Consider the example again in Fig. 6.6. The *LFS* for all parent nodes by SSTN-G1S requires, on average, 1.75 data pages (Fig. 6.6b), while SSTN-G∀S needs only 1.5 (Fig. 6.6c). This demonstrates the advantages of the SSTN-G∀S objective function.

The NP hardness of SSTN-G∀S follows from a well-known result about the NP-hardness of the following balanced hypergraph k-partitioning problem [1, 8].

Given a graph $G = (N, E)$, where N denotes a set of nodes and E a set of hyper-edges that can connect more than two nodes, the goal of the balanced hypergraph k-partitioning problem is to partition N into equal sized parts $N_1, \ldots, N_k$ while minimizing hyperedge-cuts. A hyperedge is not cut if all nodes are in one partition, and cut exactly once otherwise. This graph partitioning problem is already NP-complete for the case $k = 2$, which is also called the balanced hypergraph bi-partitioning problem [1, 8].

Theorem 5 *The SSTN-G∀S problem is NP-hard [16].*

Proof Given $STN(N, E, T)$, the SSTN-G∀S problem partitions N into equal sized parts $N_1, \ldots, N_k$ while minimizing $\sum\limits_{n \in N} |LFS(n)|$. The SSTN-G∀S problem clearly belongs to NP since given an instance of SSTN-G∀S and the maximum bound B, we can take a set of parts such that $\sum\limits_{n \in N} |LFS(n)|$ is lower than B as a valid certification. Consider the case $k = 2$. Then it is easy to show that SSTN-G∀S is equivalent to the balanced hypergraph bi-partitioning problem because $LFS(n)$ becomes a hyperedge that connects node n and its successors. Since the balanced hypergraph bi-partitioning problem is constructed from SSTN-G∀S ($k = 2$) in polynomial time, the proof is complete.

Generally speaking, the SSTN-G∀S problem is more difficult than the hypergraph partitioning problem. Consider the example in Fig. 6.7 ($k = 3$). The result of SSTN-G∀S shows that the sum of $LFSs$ is 15 (Fig. 6.7a), whereas hypergraph partitioning may yield 17 (Fig. 6.7b). As expected, SSTN-G∀S should minimize data page fragments inside LFS as well as minimize hyperedge-cuts. In the next section, we define basic concept underlying the LCP-G∀S algorithm.

6.2.2.2 Basic Concepts for LCP-G∀S

The objective function can be generalized for frequent network operations in terms of expected I/O cost. Consider the case that the weight $w(n, t)$ associated with

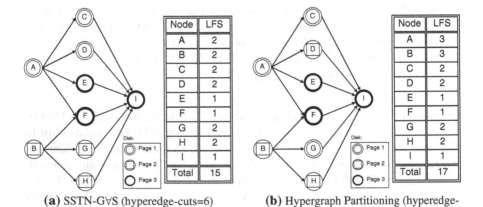

(**a**) SSTN-G∀S (hyperedge-cuts=6)   (**b**) Hypergraph Partitioning (hyperedge-cuts=6)

Fig. 6.7 Hardness of SSTN-G∀S

$Node(n, t)$ represents the relative frequency of a query accessing $Node(n, t)$. Then, every $LFS(n)$ has a weight corresponding to the access frequencies of a node n. We now generalize and formally describe our objective function that underlies the frequency of query occurrence.

$$SSTN-G\forall S(w(n)) = minimize \sum_{n \in N} w(n)|LFS(n)|, \qquad (6.2)$$

where $w(n)$ represents the relative frequency of the $LGetAllSuccessors()$ operation that accesses the node n.

Our objective for SSTN-G$\forall$S is to minimize $\sum_{n \in N} w(n)|LFS(n)|$ while preserving page-size constraints (e.g., maximum page size and minimum page utilization ratio) on each partition. One of the best-known partitioning algorithms is a Tabu-search based iterative-improvement algorithm (e.g., Kernighan-Lin(KL) and Fiduccia-Mattheyses (FM)) [6, 11]. The algorithm begins with an initial partition and iteratively moves a node n to improve the objective function. Using the idea of an FM algorithm [6], the gain (or cost) of moving a node n_1 to another partition is defined as the change in $\sum_{n \in N} w(n)|LFS(n)|$ before and after moving the node n_1.

Then, the gain of moving the node n_1 from page A to page B can be represented by

$$Gain_{A \to B}(n_1) = \sum_{n \in STN_{P(n_1)=A}} w(n)|LFS(n)| - \sum_{n \in STN_{P(n_1)=B}} w(n)|LFS(n)|, \quad (6.3)$$

where $STN_{P(n_1)=A}$ is the $STN(N, E)$ with the page-id of $n_1 = A$.

Example: Figure 6.8 shows an STN and $Gain_{2 \to 1}(E)$. The nodes on the left side are stored in Page 1 and the rest are stored in Page 2. The numbers above the nodes represent the relative frequency of a query accessing the nodes. After calling $FindNode(n)$, the additional disk I/Os for $LGetAllSuccessors(n)$ operations

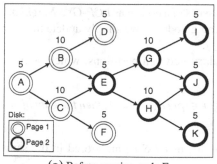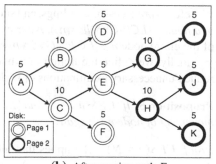

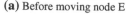

(a) Before moving node E   **(b)** After moving node E

Fig. 6.8 Move node E from Page 2 to Page 1 ($Gain_{2 \to 1}(E) = 15$)

for all nodes in Fig. 6.8b are 15 less than in Fig. 6.8a. The result of moving node E is exactly the same as the result of the gain function ($Gain_{2\to1}(E) = 95 - 80 = 15$).

The difficulty with the above gain function in practice is in computing all $|LFS(n \in STN)|$. We now show that $LFS(n_1)$ and $LFS(n \in pred(n_1))$ are sufficient to compute the gain for moving node n_1.

Proposition 3 *The gain of moving node n_1 from page A to page B is defined as:*

$$Gain_{A\to B}(n_1) = \sum_{n\in\{n_1\}\cup pred(n_1)} w(n)\big(|LFS(n, STN_{P(n_1)=A})| - |LFS(n, STN_{P(n_1)=B})|\big),$$

$$(6.4)$$

where $LFS(n, STN_{P(n_1)=A})$ is $LFS(n)$ with $STN_{P(n_1)=A}$ and $pred(n)$ is a predecessor set of a node n.

Proof If $LFS(n)$ contains a node n_1, then the node n should be n_1 or an element of $pred(n_1)$. Let $X(n_1) = \{x | x \in N, x \neq n_1, x \notin pred(n_1)\}$. Clearly, $\forall x \in X(n_1)$ have no change of their gain after node n_1 is moved because $LFS(x)$ and $LFS(n \in pred(x))$ do not contain node n_1. Therefore, we do not need to consider $LFS(n \in X(n_1))$ for the gain of node n_1.

Moving node n_1 will change other node gains. However, examining all nodes is not an efficient way to find these nodes. It is only worth examining candidate nodes that may change their $LFSs$.

Proposition 4 *The movement of node n_1 changes the gain of its predecessors, its siblings, its successors, and itself.*

Proof According to Proposition 3, the movement of node n_1 changes only $LFS(n_1)$ and $LFS(n \in pred(n_1))$. Therefore, we need to consider other node gain updates that use $LFS(n_1)$ or $LFS(n \in pred(n_1))$. Clearly, node n_1 uses both $LFS(n_1)$ and $LFS(n \in pred(n_1))$, predecessors of node n_1 use $LFS(n \in pred(n_1))$, siblings of node n_1 use $LFS(n \in pred(n_1))$, and successors of node n_1 use $LFS(n_1)$.

We refer to predecessors, siblings, and successors of a node n as *LCP-G∀S Neighbors of a node n (LCP-G∀SNbrs(n))*. After moving a node n_1, we simply update the gain of the entire node $n \in LCP-G\forall SNbrs(n_1)$.

Finally, we define the notion of a boundary between partitions, which helps to reduce unnecessary examinations of nodes.

Proposition 5 *If $|LFS(n)| = 1$ and $|LFS(m \in pred(n))| = 1$, then the gain of any movement of node n is negative.*

Proof $LFS(n \in N) = 1$ implies that all nodes in $LFS(n)$ are stored in the same data page. After moving node n, the decrease of the gain is $-(w(n) + \sum_{m\in pred(n)}$

$w(m)) < 0$.

We can divide all nodes into two parts: LCP-G∀S inside nodes ($LCP-G\forall SINs(N)$) and LCP-G∀S boundary nodes ($LCP-G\forall SBNs(N)$).

- $LCP-G\forall SINs(N) = \{n|n \in N,\ |LFS(n)| = 1\ and\ |LFS(m \in pred(n))| = 1\}$
- $LCP-G\forall SBNs(N) = \{n|n \in N,\ |LFS(n)| > 1\ or\ |LFS(m \in pred(n))| > 1\}$

LCP-G∀S boundary nodes may have positive gain or not, but LCP-G∀S inside nodes never have a positive gain. In our approach. we consider a set of LCP-G∀S boundary nodes to optimize our objective function.

Example: As can be seen in Fig. 6.8a, all nodes except nodes B, C, and E are LCP-G∀S inside nodes. Intuitively, no nodes $n \in LCP-G\forall SINs(N)$ cause additional disk I/Os for $LGetAllSuccessors()$. After moving node E to Page 1, E, G, and H become LCP-G∀S boundary nodes, as shown in Fig. 6.8b.

6.2.3 Algorithm for LCP-G∀S

Now we describe our LCP-G∀S algorithm using Tabu-search based iterative optimization. A Tabu-search explores candidate solutions and iteratively chooses the best one from a Tabu-list. It allows temporary deterioration in solution quality to escape from local optima, but eventually achieves a near-optimal solution [9, 12]. It is important to note that a high-quality initial solution reduces the overall run time of the algorithm [12]. For efficiency, LCP-G∀S uses the result of LCP-G1S (i.e., min-cut graph partitioning) as an initial solution.

An example of one pass of a two-way LCP-G∀S algorithm is shown in Fig. 6.9a. To simplify the example, we use orthogonal partitioning as an initial solution and enforce a node balancing constraint (i.e., every partition has the same number of nodes). In every step, LCP-G∀S computes gains for LCP-G∀S boundary nodes and chooses the best movement based on both the largest gain and the balancing constraint. After a node is moved, it is locked to prevent moving in the remainder of the pass. In the example, $A1$, $B1$, $C1$, $D1$, and $B2$ have the largest gain (e.g., gain = 1) at Step 1. As a tie break rule, we assume that a successor has a higher priority to move to the other page. After moving $B2$, the partitions violate the node balancing constraint. Therefore, at Step 2, $A1$ is chosen to move to the other data page due to its having the largest gain. Again, $D2$ is chosen to move to the other partition at Step 3 due to its largest positive gain. This process continues until no more positive gain takes place. Step 5 shows the partitioning result of one pass of the LCP-G∀S algorithm. As can be seen, the execution of one pass tries to minimize the SSTN-G∀S objective function from the initial solution. In contrast to LCP-G∀S, LCP-G1S relies on a min-cut gain function, as shown in Fig. 6.9b. We remind the reader that the two approaches have different objective (or gain) functions to optimize STN operations (e.g., $LGetAllSuccessors()$ and $LGetOneSuccessor()$). Even though the size of min-cuts by LCP-G∀S (Fig. 6.9a) is greater than for LCP-G1S (Fig. 6.9b), LCP-G∀S

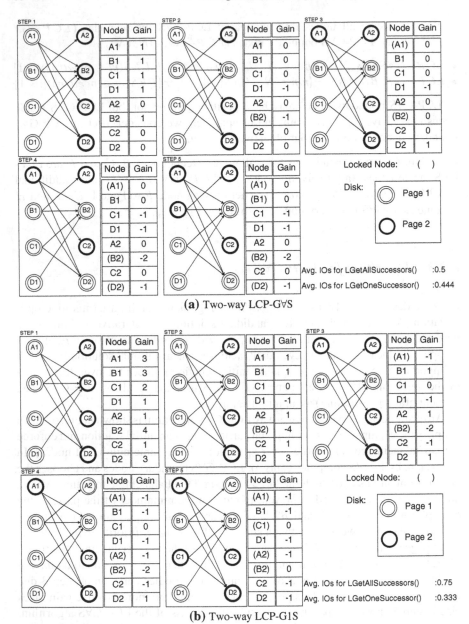

(a) Two-way LCP-G∀S

(b) Two-way LCP-G1S

Fig. 6.9 Two-way LCP-G∀S and LCP-G1S methods

shows more reduced I/O costs for the $LGetAllSuccessors()$ operation due to the smaller $\sum_{n \in N} |LFS(n)|$.

Algorithm 7 Pseudocode for the K-way LCP-G∀S algorithm

Inputs:

- LCP-G1S-STN: partitioned STN with min-cuts
- PageSizeConstraints: maximum page size and minimum page utilization ratio
- G_{min}: minimum gain threshold value

Outputs:

- Data file containing STN data across data pages (LCP-G∀S-STN)

LCP-G∀S
1:   $totalGain = \infty$
2: **while** $totalGain > G_{min}$ **do**
3:   $BN[] = LCP\text{-}G\forall SBNs$ from STN
4:   **for** each node n in $BN[]$ **do**
5:     $curPNm$ = partition number of n
6:     $NbrPNm[]$ = partition numbers of $LCP\text{-}G\forall SNbrs(n)$
7:     check PageSizeConstraints and find maxGain from NbrPNm[]
8:     add maxGain into $TB\text{-}LIST$
9:   **end for**
10:   **for** each $gain(n,curPNm,nbrPNm)$ in $TB\text{-}LIST$ **do**
11:     try moving n from $curPNm$ to $nbrPNm$
12:     lock node n
13:     update gains of $LCP\text{-}G\forall SNbrs(n)$ in TB-LIST
14:   **end for**
15:   find local maxima point and move all nodes until local maxima
16:   $totalGain$ = local maxima gain
17: **end while**

Algorithm 7 shows a way to optimize our objective function with a K-way approach. The input is an initial partition with LCP-G1S (min-cut graph partitioning) and page size constraints. The output is disk storage of the STN network. Lines 1–2 show the stop criterion when no improvement is obtained over the best solution found. Line 3 chooses possible candidates (LCP-G∀S boundary nodes) for solution modification. Lines 4–9 find the best movement for each candidate and create a Tabu-list. Lines 10–11 iteratively choose the best gain from the Tabu-list and temporarily move the node to explore a new solution. Line 12 locks the moving node to prevent revisiting the same solution in the pass. Line 13 updates the gains of $n \in LCP\text{-}G\forall SNbrs(n)$. Line 15 finds the local maxima and Lines 16, 17 repeat the process until no further improvement is possible.

6.3 Cost Models

We developed cost models for estimating disk I/Os based on the STN operations. Table 6.1 lists the symbols used to develop our cost formulas.

To determine the expected number of page accesses for $LGetAllSuccessors()$, we need to compute the distinct number of data pages across siblings. We refer to this parameter as a Sibling Collocation Efficiency (SCE). $S\hat{C}E$ denotes the average SCE in a spatio-temporal network. $S\hat{C}E$ can be obtained by one scan of a set of successors on every node along the time series. Then we integrate $S\hat{C}E$ with the $LRatio$. Let us now consider the following case: Assume that the $FindNode()$ operation retrieves a source node and that a data page for the source node is located in the buffer cache. If $LGetAllSuccessors()$ were then to retrieve all successors from the disk, the cost would be at most SCE I/Os. Then the probability that the data page of the source node is not contained in the data page set for its successors is $(1 - LRatio)^E$, resulting in a cost of one disk I/O for the $FindNode()$ operation. Therefore, the expected number of pages can now be obtained by integrating $S\hat{C}E$ and $LRatio$ as shown in Eq. (6.5).

$$\text{Cost of LGetAllSuccessors}() = S\hat{C}E + (1 - LRatio)^{\hat{E}} - 1 \qquad (6.5)$$

Table 6.2 summarizes the cost models for the STN operations. As can be seen, cost models for SSTN-G∀S and SSTN-G1S rely on the $LRatio$. However, the SSTN-G∀S cost model depends on another two parameters ($\hat{E}$ and $S\hat{C}E$). As expected, the two parameters are highly correlated to the edge/node ratio. For instance, if $\hat{E}$ becomes very large, then $(1 - LRatio)^E$ converges to 0. Therefore, $S\hat{C}E$ becomes the main factor that affects the performance of $LGetAllSuccessors()$. Consider the case where $\hat{E} = 1$; since $\hat{E}$ and $S\hat{C}E$ do not affect the performance for SSTN-G∀S (i.e., $\hat{E} = 1$ and $S\hat{C}E = 1$), the two cost models become exactly the same. Since

Table 6.1 Symbols used in cost analysis

Symbol	Meaning
$\hat{E}$	Average edge/node ratio (node degree) in STN
$LRatio$	The probability that two nodes on a Lagrangian edge are stored into the same data page
$S\hat{C}E$	Average distinct number of data pages across siblings

Table 6.2 Cost analysis for retrieval operations

Operation	Data page accesses
$FindNode()$	1
$LGetAllSuccessors()$	$(1 - LRatio)^{\hat{E}} + S\hat{C}E - 1$
$LGetOneSuccessor()$	$1 - LRatio$

SSTN-G1S is a good approximation to optimize the SSTN-G∀S objective function, we choose LCP-G1S as an initial solution for LCP-G∀S and reoptimize the partitions to minimize the SSTN-G∀S objective function.

6.4 Experimental Analysis

This section presents the experimental analysis of the LCP-G1S and LCP-G∀S algorithms are provided. The overall goal of the experiments was to show the performance improvements to access a set of successors that can be obtained by the LCP-G∀S algorithm. The metric of comparison was number of data pages accessed. We also defined a blocking factor as the number of node records per data page. We wanted to answer six questions: (1) Can our cost models predict the disk I/Os for STN operations? (2) What is the effect of the blocking factor? (3) What is the effect of the number of time steps? (4) What is the effect of the edge/node ratio in STN? (5) What is the effect of the higher edge/node ratio in real-world dataset? (6) What is the effect of node weights that can be assigned according to relative frequency of STN query?

6.4.1 Experimental Design

Figure 6.10a shows our experimental setup. We used two real datasets: (1) a Minneapolis, MN road map consisting of $1,481$ nodes and $4,574$ edges (Fig. 6.10c) and (2) a US major airline map with 226 nodes and $3,824$ edges (Fig. 6.10d). Minnesota roads are classified as either major or minor. Major roads have high traffic speeds and are often the main route to major destinations. We used real-world traffic datasets that consist of $5,760$ time steps and cover five days (Mon-Thu), obtained from the NAVTEQ corporation. To understand the effect of parameter settings for LCP-G∀S, we also synthetically increased the edge/node ratio (node degree) by adding edges according to transitive closure of the edge relation on the Minnesota road map. Major US airline routes were chosen from the USDOT dataset (http://www.transtats.bts.gov). We used 240 time steps covering five days (01/01/12 to 01/05/12).

From these STN datasets, we created and stored four database files, using four different methods: LCP-G∀S, LCP-G1S, and two orthogonal partitioning methods (snapshot and longitudinal). The stored networks were then evaluated using two operations: *LGetAllSuccessors()* and *LGetOneSuccessor()*. All experiments were performed on an Intel Core i7-2670QM CPU machine running MS Windows 7 with 8 GB of RAM. Storage and access methods were implemented based on Java 1.7 and a B+ secondary index distributed by the open-source project JDBM (http://jdbm.sourceforge.net).

Partitioning Methods: We performed experiments using four different STN storage candidates. The first two were orthogonal approaches (snapshot, longitudinal) and

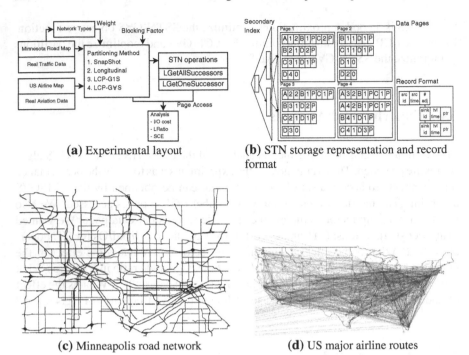

(**a**) Experimental layout

(**b**) STN storage representation and record format

(**c**) Minneapolis road network

(**d**) US major airline routes

Fig. 6.10 Experimental setup

the third and fourth were LCP-G∀S and LCP-G1S. As an initial partitioning method for both LCP-G∀S and LCP-G1S, we used Metis [10]. Then LCP-G∀S and LCP-G1S were applied to the initial partitioning to minimize their objective functions and preserve page-size constraints. Both LCP-G1S and LCP-G∀S algorithms were implemented based on Java 1.7.

STN Storage Representation and Record Format: Our STN storage model has two components: a secondary index and data pages. The secondary index (e.g., B+ tree or R tree) enables fast access to a data page using a data page pointer. A data page stores records using an adjacent record format. A record has a source-node id, a source-node time, and the number of incidents. Every incident has a node-id, a travel time, and a data page pointer for the incident record. For example, node $A1$ has two incidents (B, C) and occupies 9 record units in Fig. 6.10b. In our analysis, the default minimum page-space utilization ratio was 50% and the average page-space utilization ratio was 70%.

6.4.2 Experimental Results and Analysis

Evaluation of Cost Model: The aim of the first set of experiments was to demonstrate the accuracy of our cost model given in Sect. 6.3. We used a uniform weight (i.e., all weights on nodes = 1) to simplify the interpretation of the results. The experiments used a blocking factor of 8 and chose randomly 50% of the total number of nodes. Table 6.3 summarizes the real and predicted disk I/Os of the STN operations with two real-world STN datasets. In our experiment, we used one buffer cache to store only one data page in memory and called a $FindNode()$ operation before calling a $LGetAllSuccessors()$ or a $LGetOneSuccessor()$. As the table shows, the prediction error for our cost models is within 1%. When we called $LGetAllSuccessors()$, LCP-G∀S outperformed the other approaches. By contrast, when we called $LGetOneSuccessor()$, LCP-G1S performed best.

Effect of Blocking Factor: The second experiment evaluated the effect of the blocking factor on the performance of the algorithms. To evaluate the performance of alternative access methods, we worked with two STN datasets: one was a real Minnesota road map and the other, a synthetic road map with an increased edge/node ratio. We fixed a query set and increased the blocking factor. As shown in Fig. 6.11a,b, we observed an improvement in disk I/O efficiency for the two LCP methods over orthogonal methods, as expected, due to its ability to store temporally connected

Table 6.3 The I/O costs of STN operations

Operation								
Method	LGetAllSuccessors()			LGetOneSuccessor()			LRatio	SCE
	Actual	Predicted	Pred. error (%)	Actual	Predicted	Pred. error (%)		
LCP-G∀S	1.5236	1.5150	0.57	0.6718	0.6718	0	0.3282	2.2222
LCP-G1S	1.7189	1.7192	0.02	0.5884	0.5886	0.04	0.4114	2.5245
SnapShot	3.0466	3.0470	0.01	1	1	0	0	3.0470
Longitudinal	3.0469	3.0474	0.01	1	1	0	0	3.0474

Road traffic dataset with time steps $=1{,}440$, $\hat{E}=3.0885$, and blocking factor$=8$

Operation								
Method	LGetAllSuccessors()			LGetOneSuccessor()			LRatio	SCE
	Actual	Predicted	Pred. error (%)	Actual	Predicted	Pred. error (%)		
LCP-G∀S	10.2470	10.2932	0.44	0.9788	0.9788	0	0.0212	10.6058
LCP-G1S	13.9845	14.0956	0.78	0.9439	0.9438	0	0.0562	14.7320
SnapShot	16.4232	16.5919	1.01	1	1	0	0	16.5919
Longitudinal	16.8846	17.0747	1.11	1	1	0	0	17.0747

Aviation dataset with time steps $=240$, $\hat{E}=17.4898$, and blocking factor$=8$

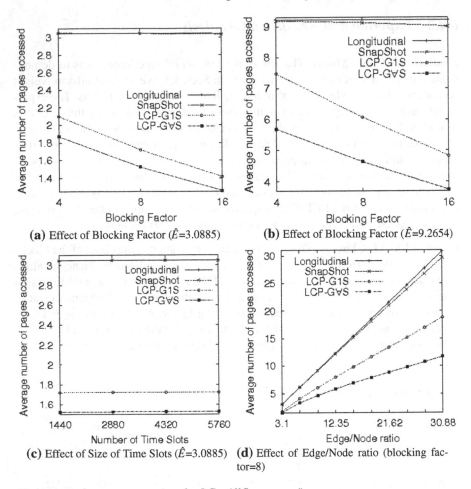

(a) Effect of Blocking Factor ($\hat{E}$=3.0885)   **(b)** Effect of Blocking Factor ($\hat{E}$=9.2654)

(c) Effect of Size of Time Slots ($\hat{E}$=3.0885)   **(d)** Effect of Edge/Node ratio (blocking factor=8)

Fig. 6.11 Performance comparison for *LGetAllSuccessors()*

information on a single data page. A larger blocking factor enhanced disk I/O perfor-mance for both LCP-G∀S and LCP-G1S. This is because more data will be collocated based on Lagrangian-connectivity as the blocking factor increases. By contrast, we observed that longitudinal and snapshot showed no difference in disk I/Os.

Effect of Number of Time Steps: The third experiment evaluated the effect of the length of time series on the performance of the algorithms. We increased the number of time steps and therefore increased the number of data pages proportional to the increased number of records. The effect of the length of time steps on I/O costs is shown in Fig. 6.11c. As can be seen, time series length does not affect performance of any method. This property is desirable when a user incrementally stores large numbers of time steps because it allows us to carve a large number of time steps into tiny sections and store them individually.

Effect of Edge/Node Ratio: The fourth experiment evaluated the effect of the edge/node ratio on network datasets. We synthetically connected nodes based on the transitive closure of a spatio-temporal directed graph and then increased the edge/node ratio. Figure 6.11d shows that a higher edge/node ratio increases the performance gap between LCP-G∀S and LCP-G1S, with LCP-G∀S performing better than LCP-G1S. These results demonstrate that the performance of LCP-G∀S relies on both the *LRatio* and $\hat{E}$, as shown in Table 6.2.

Aviation STN Dataset: The fifth experiment evaluated the performance of the LCP-G∀S algorithm with a real-world aviation STN dataset. First, we chose 266 major airports, obtained from USDOT (http://www.transtats.bts.gov). Then, we refined this selection to include more connected airports (120 airports) in terms of the number of flights between them. Figure 6.12a shows that LCP-G∀S reduced 27% disk I/Os for *LGetAllSuccessors*() over LCP-G1S. When considering more connected STNs (Fig. 6.12b), the performance gap between LCP-G∀S and LCP-G1S increases up to 30%.

Effect of Node Weight: The sixth experiment evaluated the effect of weights on the LCP-G∀S algorithm. First, we assigned a specific weight on major roads and used a unit weight (i.e., $w(n) = 1$) on minor roads. Then, query sets were chosen from the two types of roads according to an access frequency (weight). Figure 6.13a,b show that weighted LCP-G∀S decreases the I/O costs in the case of a given relative frequency. When considering $w(n) = 4$ for major roads and $w(n) = 1$ for minor roads, the Weighted LCP-G∀S reduced 9% disk I/Os for *LGetAllSuccessors*() over non-weighted LCP-G∀S.

In our experimental analysis, LCP-G∀S showed the best performance on the *LGetAllSuccessors*() operation against other approaches. Although the *LRatio* was a good approximation to optimize the operation, it showed a limitation to achieve

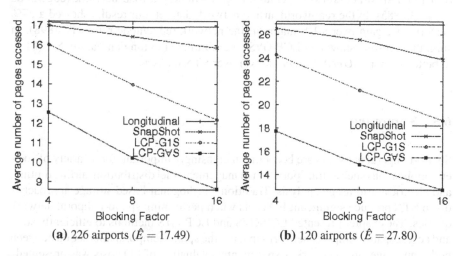

(a) 226 airports ($\hat{E} = 17.49$)   (b) 120 airports ($\hat{E} = 27.80$)

Fig. 6.12 Performance comparison with real aviation STN dataset

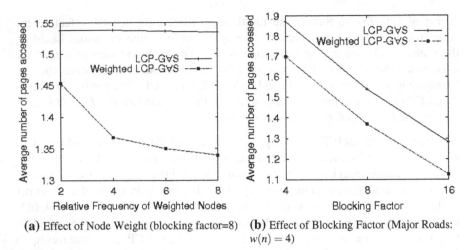

(a) Effect of Node Weight (blocking factor=8)   (b) Effect of Blocking Factor (Major Roads: $w(n) = 4$)

Fig. 6.13 Performance comparison between LCP-G∀S and weighted LCP-G∀S

the optimal partitioning. We defined the SSTN-G∀S objective function and showed a way to optimize the function according to our LCP-G∀S algorithm. In our I/O analysis, we introduced the notion of Sibling Collocation Efficiency (SCE) and formulated a cost model based on both *LRatio* and SCE. Because a higher edge/node ratio has more chance to minimize SCE, LCP-G∀S outperforms LCP-G1S when the edge/node ratio increases. When the edge/node ratio was 1, there was no difference between LCP-G∀S and LCP-G1S. When the edge/node ratio was 30.88, LCP-G∀S showed a 38% disk I/O reduction over LCP-G1S. In real-world road networks, the edge/node ratio was close to 3.0. Our results showed a 12% performance gain of a $LGetAllSuccessors()$ operation over a Minnesota road network (edge/node ratio=3.0885). In the real-world airline network dataset, our results showed a 27% performance gain over a US major airline network (edge/node ratio=17.4898). In sum, these results show that LCP-G∀S can minimize Lagrangian Family Set (LFS) to better support $LGetAllSuccessors()$ over STN datasets.

6.5   Summary

Spatio-temporal networks are becoming increasingly important for a variety of societal applications such as transportation management, fuel distribution, airline routing, and electrical grid usage analysis. Traditional orthogonal-based storage approaches on an STN produce significant I/O costs when performing spatio-temporal network queries. The chapter presented LCP-G∀S and LCP-G1S methods to efficiently store and access spatio-temporal networks that use the spatio-temporal interaction between nodes and edges in a network. Experimental evaluation of I/O costs was presented.

References

1. Alpert C, Kahng A (1995) Recent directions in netlist partitioning: a survey. Integr VLSI J 19(1–2):1–81
2. Batchelor G (2000) An introduction to fluid dynamics. Cambridge University Press, Cambridge
3. Evans MR, Yang K, Kang JM, Shekhar S (2010) A Lagrangian approach for storage of spatio-temporal network datasets: a summary of results. In: Proceedings of the 18th SIGSPATIAL international conference on advances in geographic information systems, pp 212–221, ACM
4. Evans MR, Yang K, Kang JM, Shekhar S (2010) A Lagrangian approach for storage of spatio-temporal network datasets: a summary of results. In: GIS, pp 212–221
5. Evans MR, Yang K, Gunturi V, George B, Shekhar S (2015) Spatio-temporal networks: modeling, storing, and querying temporally-detailed roadmaps. In: Space-time integration in geography and GIScience. Springer, Dordrecht, pp 77–108
6. Fiduccia C, Mattheyses R (1982) A linear-time heuristic for improving network partitions. In: 19th conference on design automation. IEEE, pp 175–181
7. Ford LR, Fulkerson DR (1958) Constructing maximal dynamic flows from static flows. Oper Res 6(3):419–433 May–June 1958
8. Garey MR, Johnson DS (1990) Computers and intractability; a guide to the theory of NP-completeness. W.H. Freeman & Co., New York
9. Glover F (1990) Tabu search-part II. ORSA J Comput 2(1):4–32
10. Karypis G, Kumar V (1999) A fast and high quality multilevel scheme for partitioning irregular graphs. SIAM J Sci Comput 20(1):359
11. Kernighan B, Lin S (1970) An efficient heuristic procedure for partitioning graphs. Bell Syst Tech J 49(2):291–307
12. Michiels W, Aarts E, Korst J (2007) Theoretical aspects of local search. Springer, New York Inc
13. Pallottino S, Scutella M (1998) Shortest path algorithms in transportation models: classical and innovative aspects. In: Equilibrium and advanced transportation modelling, pp 245–281
14. Shekhar S, Liu D (1997) CCAM: a connectivity-clustered access method for networks and network computations. IEEE Trans Know Data Eng 9(1):102–119
15. Turner S, Margiotta R, Lomax T (2004) Lessons learned: monitoring highway congestion and reliability using archived traffic detector data. US Department of Transportation, Federal Highway Administration
16. Yang K, Evans MR, Gunturi VM, Kang JM, Shekhar S (2014) Lagrangian approaches to storage of spatio-temporal network datasets. IEEE Trans Know Data Eng 26(9):2222–2236

Chapter 7
Summary

Spatio-temporal networks (STN) are today the very heart of our modern infrastructure. Ever increasing volumes of STN data collected from GPS systems, sensors on roadways and in vehicle engines, etc. must be stored and managed to support transportation safety, fuel distribution, electrical grid systems, airline routing, disease tracking, and a myriad of other applications. Current spatial computing techniques and database systems are inadequate for the task. The size and continual updating of spatial network big databases make it difficult to meet the most important requirements of any database: easy and timely accessibility by users. New computing techniques and database systems are needed to handle the special challenges posed by spatial network big data.

Developing Spatial Network Big Database Systems requires overcoming three key challenges. First, it requires a new data model to represent the complex and interrelated structure of SNBD. Second, fully exploiting SNBD requires scalable query processing and optimization methods, which are currently lacking. Finally, SNBD requires I/O efficient storage and access methods that leverage scalability and efficiency of big data query processing. To address the challenge of data modeling for SNBD, we reviewed multiple ways to represent SNDB datasets and explored SNBD models based on time-aggregation graphs and spatio-temporal constraints. To address the challenges of scalable query processing and optimization methods, we explored computational techniques that can utilize the properties of SNBD in order to develop scalable query processing algorithms. We studied three application domains and applied query optimization techniques to improve the performance time for query processing. First, we studied query processing for resource and shelter allocation in the wake of human-made and natural disasters and explored algorithms for the Capacity-Constrained Network-Voronoi Diagram (CCNVD) problem. Second, we studied query processing for spatial coverage planning and optimization for accident events and explored algorithms for the Distance-Constrained k Spatial Sub-networks problem. Finally, we studied query processing for reducing traffic congestion during or after disasters and explored algorithms for the Evacuation Route Planning (ERP)

© Springer International Publishing AG 2017

K. Yang and S. Shekhar, *Spatial Network Big Databases*,

DOI 10.1007/978-3-319-56657-3_7

problem. To address the challenges of I/O efficient storage and access methods, we explored data partitioning algorithms for spatio-temporal networks and optimized two data access operators: *GetASuccessor*() and *GetAllSuccessors*().

This book presented four SNBD problem formulations and their solutions. We end by summarizing the contributions these works make to effectively harness the power of spatial network big databases. The contributions are organized according to the challenges they help address.

7.1   Capacity Constrained Network Voronoi Diagram

We described a Pressure Equalizer (PE) approach for creating a CCNVD that meets the capacity constraints of service centers while maintaining the contiguity of service areas assigned to those centers. The core idea in PE is the Pressure Equalization Graph that can balance the workload of each service center under the service area contiguity constraint. However, improving PE's scalability to large sized transportation networks requires addressing the computational bottleneck that occurs during Service Area Contiguity Checking. To remedy this problem, we introduced two main ideas: (1) the Block Tree Contiguity Checking method that can efficiently test the contiguity of service area and (2) the Graph-Minor method that can reduce the size of the initial solution. We showed that Block Tree Contiguity Checking coupled with the Graph-Minor method can successfully reduce the computational cost of the PE algorithm.

7.2   Distance-Constrained k Spatial Sub-networks

We introduced the notion of the rooted sub-graphs that can construct nearest neighbor distribution (NND) functions in the spatial network. The RSG-NND algorithm solves the DCSSN problem using an iterative approach to search an attractor node and construct a DCSSN based on attractor nodes and NND functions. We demonstrated how an approach based on RSG-NND can estimate the density of spatial events and create k spatial sub-networks.

7.3   Evacuation Route Planning

Algorithms for evacuation planning must be able to account for the capacity constraints of the road network with manageable computational cost. In this problem, we introduced a dartboard network structure to reflect an evacuee flow pattern for common evacuation scenarios by exploiting the spatial structure of the road network.

The DBNC-ERP algorithm partitions the network using dartboard network cuts (DBN-cuts) and groups source nodes in different spatial locations to maximize the number of evacuees.

7.4   Storage Schemes for Spatio-Temporal Network Datasets

Traditional orthogonal-based storage approaches incur significant I/O costs when performing spatio-temporal network queries. We presented LCP-G∀S and LCP-G1S methods to efficiently store and access spatio-temporal networks that use the spatio-temporal interaction between nodes and edges in a network. The core idea of these two approaches is to conceptualize the Lagrangian connectivity in STN data models and partition the STN data to minimize I/O costs for data retrieval operations.

Printed in the United States
By Bookmasters